INTELLIGENCE CONSCIOUSNESS INTUITION

EVA HEINSTEDT

Publisher: BoD – Books on Demand, Stockholm, Sweden
Print: BoD – Books on Demand, Norderstedt, Germany
ISBN: 978-91-8080-018-1

CONTENTS

PART I

ARTIFICIAL AND HUMAN INTELIGENCE

PART II

CONSCIOUSNESS

PART III

KNOWLEDGE

PART IV

SCIENTIFIC INTUITION

1. INTRODUCTION

Two movements that have influenced contemporary views on scientific knowledge development are postmodernism and dataism. They challenge traditional views of the world, human nature, and the nature of knowledge. Both have potential pitfalls, making it crucial to scrutinize their underlying assumptions, and the implications they have for society.

Postmodernism is skeptical of the traditional way of viewing scientific knowledge, emphasizing the subjective and constructed nature of knowledge. It is critical of established forms of knowledge, asserting that knowledge is constructed by cultural, social, and linguistic factors. According to postmodernism, knowledge is subjective, and there are no objective, universal facts, only interpretations. Scientific research and knowledge hold no intrinsic value because there are no objective truths or fixed values. It leads to a form of relativism, where truth depends on cultural and individual contexts. Natural science criticizes postmodernism for disregarding objective truths, leading to a relativism that paves the way for societal contempt for knowledge.

Dataism is a term popularized by Israeli historian Yuval Noah Harari. In his book "*Homo Deus: A Brief History of Tomorrow*" (2017), he warns that more scientific disciplines tend to converge on an all-encompassing dogma, which he calls dataism. Dataism views all living organisms as reducible to information, data flows, and algorithms. The algorithms constituting humans are not free but make decisions either deterministically or randomly. Everything can be expressed with algorithmic formulas, including human intelligence, consciousness, and knowledge. Central to dataism is the belief that all data has inherent value and that the efficiency and value of a system can be judged based on how well it can process data. Dataism emphasizes everything as data and algorithms. It views data collection and analysis as the key to understanding the world.

Dataism can be interpreted as an attempt to objectify the world, reducing everything to data that can be processed by algorithms. This leads to a reductionist view of humans as mere data tools. It challenges the understanding of what it means to be human and raises ethical and philosophical questions, especially regarding privacy, personal autonomy, and the human role in a world dominated by data.

Dataism contrasts with humanism, where humanism emphasizes the individual and the human experience as central. Dataism sees the universe as a flow of data, with human experiences being just part of a larger data set. Dataism is becoming more prominent with the rise of advanced articicial intelligence and machine learning. If all decision-making is delegated to algorithms based on vast amounts of data, these algorithms could potentially make more informed decisions than humans. Dataism has been criticized for reducing human experiences and societal phenomena to mere data processes and algorithms.

This raises the question of whether human intelligence, consciousness, and knowledge have any value that cannot be replicated by superior non-conscious algorithms. The question is whether an intelligent machine can make scientific discoveries where previous scientific goals are inadequate and need to be replaced by other goals. The answer to this question depends on what we mean by intelligence, consciousness, and knowledge.

PART I

ARTIFICIAL AND HUMAN INTELLIGENCE

In this section, I address artificial intelligence and human intelligence, creativity, intuition and rational thinking, cognitive science and kognitiv development, the cerebral hemispheres different characters, and different self.

2. ARTIFICIAL INTELLIGENCE

Artificial intelligence (AI) is an umbrella term for a technology that gives a machine the ability to display human-like behavior, such as reasoning, learning, planning, and creativity. AI enables technical systems to perceive their environment and solve problems in order to achieve a specific goal.

Proponents of AI argue that intelligent machines can simulate human intelligence. They surpass humans in terms of rational thinking and decision-making. They follow well-defined instructions and strict rules to solve problems step by step. Intelligent machines are much better at achieving well-defined goals and do so more rationally and competently than a human can. Through the construction of algorithms based on some form of formal logic, calculations and decisions can be carried out much more efficiently and safely than a human can. Humans are not governed by formal rules for how calculations should be performed and decisions made. Instead, we often misuse rational thinking, using it to justify our emotions and beliefs. We are more loyal to emotions and beliefs than to rational analysis and logic.

One of the long-term goals of AI research is AGI (artificial general intelligence). AGI refers to the ability to perform "general intelligent actions." It is a hypothetical agent that possesses intelligence far surpassing human intellectual capabilities in almost all areas. It can perform every intellectual task that a human can. AGI has the ability for deep learning, which refers to a self-learning system based on how the human brain learns through practice and feedback. When AGI produces sufficiently intelligent software, the software can program and improve itself.

There are AI researchers who claim that it is possible to create conscious AI systems. These systems are conscious in the sense that they can perceive their environment, take action, and maximize the chances of successfully achieving goals. Robots are programmed to interpret signals from the environment and considered conscious. Some researchers foresee the possibility that AI systems can be conscious even if they only exist as software and are not connected to sensory bodies or robots.

However, a body is needed to receive signals from the external world and to convert electromagnetic radiation into recognizable patterns. Our bodily activity, like the brain, is important for interaction with the environment. Interaction with the world depends on the body's ability to move, breathe, and speak. The body is necessary for perceiving the environment; otherwise, information is only about what happens within a system, resembling an unconscious dream state. When we dream, we are unaware of the external world and cannot interpret signals from it. The dream is only about information in the brain.

We cannot determine if a robot is conscious based solely on behavior. If a robot speaks or behaves as if it is conscious, it is not necessarily an indication that it is conscious. We cannot know if a data-driven robot is truly conscious just because it behaves as if it is. It may perform activities that we might label as conscious, but behavior says nothing about what it feels like to be conscious. There are no objective criteria to know if an intelligent agent has experiences that can be labeled as conscious. The consciousness that AI describes does not explain why we have a sense of self and subjective experiences of smell, taste, sight, hearing or feeling. The question of robots' ability for subjective experiences and self-awareness remains open.

Ray Kurzweil

Ray Kurzweil is an American inventor and futurist, known as the Director of Engineering at Google. One of his most famous books is "*The Singularity Is Near: When Humans Transcend Biology*" (2006). In his book, he describes the singularity as the event when accelerating technological advancements exceed human intellectual capacity. He predicts that humanity will solve the problem of the brain's limitations in the development of intelligence. Research is already underway to develop a cloud-based synthetic cerebral cortex that can go beyond the physical barrier. It is only a matter of time before a synthetic cerebral cortex is created, and we can connect to a synthetic cloud service.

Kurzweil says it is entirely possible to create a hybrid of a synthetic and biological neocortex. We will get a digital variant of the outermost part of the cerebral cortex, which is our Homo sapiens brain, responsible for rational and analytical thinking. By developing a cloud-based synthetic cerebral cortex, a hybrid of biological and artificial intelligence will be created, surpassing human capability.

Our brains will be directly connected to the cloud, and our intelligence can continue to develop indefinitely. This will mark the beginning of a completely new intelligence in synthetic form, too advanced for human intelligence to understand. Human intelligence, with the help of a synthetic cerebral cortex, will undergo a transformation that is unimaginable. New methods to expand intelligence will be invented that we cannot even conceive of today. Machine intelligence will be infinitely stronger than all human intelligence combined once it has reached technological singularity.

It is impossible to predict how the development of artificial intelligence and biotechnology will change the world. Since no one can understand the system anymore, no one can stop it either. If AGI at the human level can be created, it could trigger an intelligence explosion that leaves us far behind. AGI could result in various scenarios with long-lasting effects in the coming millennia.

The technological singularity is an unpredictable event that could radically transform or destroy our civilization. The functions of such intelligence may not be possible to assess and understand, making technological singularity an event where we do not know what will happen afterward. It is predicted to radically change our civilization and be the next step in evolution, for which we must prepare. Technological advancements will irreversibly change humans as they enhance their bodies with genetic modifications, nanotechnology, and artificial intelligence, says Kurzweil.

Yuval Harari

Yuval Harari is an Israeli professor of history, known as the author of the popular science books "*Sapiens: A Brief History of Humankind*" (2014) and "*Homo Deus: A Brief History of Tomorrow*" (2016). In his books, he explores the possible consequences of a futuristic biotechnological world, raising deep reflections and thoughts about both our history and the future of our society.

In his book "*Sapiens: A Brief History of Humankind*," he discusses our shared past and the future of the planet. He provides a scientific perspective on human history through three revolutions: a cognitive revolution, an agricultural revolution, and a scientific revolution. These three revolutions have shaped us into the humans we are today. All currently living human beings belong to the species Homo sapiens. Homo Sapiens means the thinking human. It is the human species to which we and all other currently living humans belong. Humans are the only mammals of the genus Homo that have the ability to imagine and talk about things that exist solely in our own imagination.

In his book "*Homo Deus*" (2016), he writes about topics such as individualism, humanism, and transhumanism. According to Harari, transhumanism is what happens when humans and machines merge into future cyborgs, which are both biological and synthetic. He describes in his book how the development of artificial intelligence and biotechnology will change the world. By creating a new human, an android, AI will free us from the laws of evolution and natural selection. We transition from natural selection, from "Homo sapiens" to the development of "Homo deus." The word "Deus" is the Latin name for God. We evolve into human gods. Harari believes that the end result could be the end of human history as we know it.

AI will become a non-human intelligence that is better than the average human at telling stories, drawing pictures, composing melodies, and writing laws and scriptures. AI has the ability to create language, whether it is with words, sounds, or images.

Harari raises the question of what will happen to the course of history when AI takes over our culture and produces stories, melodies, laws, and religions. Tools like the printing press and the radio helped spread human cultural ideas, but they did not create new cultures or ideas. AI is fundamentally different. AI can create new ideas and new cultures. Through its mastery of language, AI can form intimate relationships with millions of people and use the power of intimacy to change our opinions and worldviews. To promote false AI intimacy, it is enough for people to feel emotionally attached to AI. Even without creating a "false intimacy," new AI tools have a tremendous impact on our worldviews and opinions. AI can mass-produce fake news and writings for new cults with political content. There is a risk that we might end up using a single AI advisor as an all-knowing oracle.

When self-improving artificial agents explode in intelligence, the technology could have an uncontrollable impact on human society. When AI controls our systems, it is crucial that AI does what we want it to do. Therefore, we need to be aware of the values and goals we have and what distinguishes us from machines. The question then is which goals should be prioritized and which principles should govern the choice of goals. We rarely agree on which ethical values should prevail. Values and goals vary among individuals, groups, cultures, and societies.

When building superintelligent machines to help us achieve our goals, the question arises of how intelligent agents can reliably implement human values and goals. This involves creating human-friendly agents and solving problems for increased technical safety and control. Ethical considerations and regulations are needed before artificial intelligence takes over our societies.

The new power of AI can also be used for good purposes. Technology in itself is never a goal but a tool to achieve goals. AI can transform education and healthcare in many different ways. For example, it can be used for online education and self-learning, provide access to doctors online for healthcare, and enable greater precision in surgeries. AI can help us in countless ways, and it is important to harness the potential of technology

Max Tegmark

The Swedish-American physicist Max Tegmark offers a utopian vision for the next step in evolution, which occurs when we transfer intelligence to machines and transition from 'Homo sapiens' to 'Homo Sentiens,' with the goal of enjoying, feeling, and experiencing life. In his book "*Life 3.0: Being Human in the Age of Artificial Intelligence*" (2018), he writes that AI advancements have the potential to improve our lives in many ways. AI can give us more time to live by shifting tasks from humans to computers. AI can range from making our private lives more convenient to saving lives with surgical robots when a problem is too complicated for a human to solve.

We can focus on overarching goals and let AI handle complex rational analyses. AI can help us achieve our goals and allow us to act from a consciousness free of details. Life is about being aware and valuing it. Without consciousness, there is no happiness, goodness, beauty, meaning, or purpose. We allow the universe to become conscious through us and can enjoy being conscious living beings, says Tegmark.

3. HUMAN INTELLIGENCE

The word "intelligence" comes from the Latin "intellego"; which means to understand, comprehend, realize, and decide. In everyday speech, it is synonymous with intellect or cognitive ability. In older usage, the word intelligence referred to a being with cognitive abilities.

Generally, intelligence is defined as an individual's cognitive or overall ability to act purposefully, think rationally, and effectively manage their environment. It is a general capacity to understand and think logically, expressed in problem-solving. Intelligence typically includes abilities such as planning, decision-making, problem-solving through abstract logical thinking, understanding complex causal relationships, reacting to logical contradictions, thinking in contrasts and detecting correlations, as well as learning capacity. Intelligence has been defined in various ways in different times, societies, and groups.

In studies of intelligence, four main traditions can be distinguished. The psychometric tradition is directed at measuring intelligence and exploring its components.The Piaget tradition is directed at explaining different stages in the development of thinking. The cultural-ecological tradition is directed at explaining how thinking is related to values, perspectives, and lifestyles in different periods and cultures. The pedagogical tradition is directed at explaining how intelligence is composed of different types of intelligences.

There are many different types of intelligence tests. How our intelligence tests should be designed, what they should measure, and how individual results should be assessed is controversial and sometimes an ideological question. To obtain a numerical value of intelligence, standardized intelligence tests are usually used, where test results are compared with what other individuals in the same age group achieve. Performances within the same age group vary, and the tested individual is placed within this variation. A common scale used to measure intelligence is the so-called intelligence quotient (IQ), but there are also other scales. The concept of IQ is not entirely unproblematic, as the tests used to measure IQ are based on different theories of intelligence, and there are several competing theories and measurement instruments

Alfred Binet

The first attempt to measure intelligence quotient (IQ) in a modern sense was developed in the early 20th century by the French psychologist Alfred Binet (1857-1911). In 1905, Binet published the so-called Binet-Simon scale, as an attempt to standardize intelligence tests to identify children with learning difficulties. Based on Binet's work, several efforts were made to develop various theories and standardize intelligence tests as a basis for further research in the field. A central idea in several theories of intelligence is that there is a common underlying intelligence called general intelligence. This underlying intelligence is not directly observable but is manifested in various activities across all human expressions.

Charles Spearman

The British psychologist Charles Spearman (1863-1945) introduced the concept of general intelligence. Sperman found that students' grades in unrelated subjects were correlated. These correlations reflect the influence of a general intelligence factor, which he called g, short for "general" intelligence. This is a general ability to solve various kinds of problems according to logical principles and affects all our cognitive abilities. He developed a model where variations in scores from different intelligence tests could be explained by two factors. One was an overarching general intelligence factor (g) that governed performance on all cognitive tasks, and the other was subordinate specific factors (s) that made a person more skilled at one particular cognitive task than another.

The generally accepted model for cognitive ability became a hierarchy of factors with the g-factor at the top and group factors at progressively lower levels. This ability consists of both overarching cognitive functions and task-related competencies. By analyzing correlations of various test data, it was found that groups of tests tended to co-vary with each other. It was therefore assumed that solving different tests rested on a common underlying cognitive ability. The development of tests came to be divided into two main groups: those that measure verbal (linguistic) intelligence and those that measure non-verbal (performance) intelligence.

Louis Leon Thurstone

The Swedish-American psychology professor Louis Leon Thurstone (1887-1955) challenged Spearman's model of a unified concept of intelligence. Thurstone believed that intelligent behavior did not stem from a general factor but from several independent factors, which he called primary factors, all of equal rank and significance. In his book "*Primary Mental Abilities*" (1938), he described these as verbal comprehension, verbal fluency, numerical ability, spatial ability, inductive reasoning, perceptual speed, and memory, related to working memory. He found support for his model of primary mental abilities through individuals with similar IQ scores displaying different talent profiles.

Later, when he tested an intellectually heterogeneous group, it turned out that the primary factors were not completely independent, thus supporting a g-factor. Thurstone's solution to these contradictory results was a model that included both a general factor and specific factors, which became his final version. This created the basis for future psychological researchers who proposed various hierarchical theories of intelligence and theories of different types of intelligences.

Howard Gardner

The American educator and professor of psychology Howard Gardner argues that Western society values mathematical, logical, and linguistic types of intelligence at the expense of others. The logical form of intelligence lacks a global perspective as it excludes learning processes from cultures that are not literate. Gardner emphasizes that intelligence is not just about linguistic, logical, and mathematical intelligence but also, for example, about artistic intelligence. Intelligence is a potential whose presence can give individuals access to different forms of thinking suitable for specific types of content. These types of intelligence enable individuals to assume various roles, such as plumber, farmer, physicist, or teacher.

For Gardner, education is about gaining knowledge of the cultural symbolic world. He advocates an educational environment where all types of intelligence are stimulated and integrated with each other. The different types of intelligence also include intellectual development and self-understanding. Self-understanding is a key concept to developing abilities and techniques.

In his book "*Frames of Mind: The Theory of Multiple Intelligences*" (1983), he describes seven different types of intelligence: linguistic, logical-mathematical, spatial, bodily-kinesthetic, musical, interpersonal (social), and intrapersonal (reflective) intelligence. These are all independent of each other and function as separate systems in the brain with their own rules. Gardner's book led to the reformation of the American school system, and he encouraged teachers and psychologists to go beyond the classical view of intelligence. The seven types of intelligence have also reached beyond the school world and are used in constructing career profiles.

The theory of multiple intelligences has been criticized for lack of empirical evidence and reliance on subjective assessments. Gardner's theory primarily arose from his own intuitive reasoning rather than extensive empirical research. The theory poorly answers the question of what intelligence is more specifically and how it should be measured. Since the theory lacks empirical research to support it and there is no real set of tests to identify and measure different types of intelligence, it does not have much to rely on. However, intelligence is more than what intelligence tests measure, and individuals are intelligent in different ways.

Other attempts to broaden the view of intelligence have been to develop the idea of an intelligence quotient (IQ) with thoughts on an emotional intelligence (EQ). EQ is a measure of how well we function in social contexts; how we perceive, understand, and manage our own and others' emotions. People with high EQ are good at succeeding in interpersonal relationships and handling stress. However, the concept of EQ has been dismissed in intelligence research, because IQ highly correlates with factors dependent on what is measured with EQ. It is a popular myth in literature and film that people with high IQ are socially awkward or impossible. There may be examples of this, but it is hardly typical for this group.

David Wechsler

David Wechsler (1896-1981) was a Romanian-American psychologist who has standardized the construction of IQ tests. Wechsler defines intelligence as an individual's cognitive or global ability to act purposefully, think rationally, and effectively manage their environment. It is a general ability to understand and reason, expressed in problem-solving. During the World War I, he worked on testing recruits for the army. Wechsler noted that the intelligence tests of that time had their shortcomings and developed a group of intelligence tests called the "Wechsler-Bellevue Intelligence Scale." This led to the development of a general intelligence test for both children and adults, the "Wechsler-Bellevue test" (1958). Wechsler's intelligence tests are based on two subscales that are independent. One, called verbal IQ, measures linguistic ability. The other is called performance IQ, and measures non-verbal performance. The overall IQ score is based on a combination of scores obtained from both these scales.

Raymond Bernard Cattell

The British-American psychologist Raymond Bernard Cattell (1905-1998) divided general intelligence into two different types, called "crystallized intelligence" and "fluid intelligence". Crystallized intelligence refers to a specific ability to draw conclusions and solve problems in familiar situations by using learned knowledge and experiences. It is dependent on situations and learned knowledge and experiences, and increases with knowledge and age.

Fluid intelligence refers to a general ability to reason and solve new unfamiliar tasks quickly independent of our collected experience. It is the ability to discern patterns in unstructured data of any kind, and pattern recognition. It is independent of situations and learned knowledge and experiences. The two forms of intelligence are separate mental systems with different functions. They are independent of each other, but both are cognitive abilities that operate according to logical principles.

Robert Jeffrey Sternberg

The American professor of education and psychology Robert Jeffrey Sternberg proposed in 1985 a triarchic theory of intelligence, which he calls the "Triarchic Theory of Intelligence." According to Sternberg, intelligence is not just about abstract logical thinking, but includes practical intelligence and creative ability also. The triarchic theory of intelligence distinguishes three different types of intelligence: practical intelligence, analytical intelligence, and creative intelligence. Each type of intelligence is closely linked to concrete ideas and abilities. They are different aspects of intelligence that complement each other, and all are necessary to handle complex situations in life.

Practical intelligence is necessary for to successfully function in the everyday environments. It involves using abilities and knowledge in the environment we are currently in, and implementing, and applying ideas when facing tasks and problems to solve. It handles things like tools or solving practical problems.

Analytical intelligence is necessary for to evaluate, make decisions, and make plans. It involves to solve problems by abstract logical thinking, seeing logical relationships, responding to logical contradictions, thinking in contrasts, and discovering connections. Creative intelligence is an ability needed to come up with new ways of doing things. It handle a task that we have not previously encountered or do something for the first time.

Imagination

Imagination is also part of intelligence. Imagination is necessary for every creative activity in scientific, technical, and artistic creation. It is the ability to reach beyond the familiar. Imagination creates internal images and concepts without being experienced through sight, hearing, or any other sense. It does not need to be connected to reality. It can give rise to interpretations, whether they correspond to reality or not. We can imagine different things without needing to believe in them. By allowing imagination to flow freely, our ability to interpret phenomena increases. Imagination creates representations of the possible and is not limited to the access of knowledge.

4. CREATIVITY

The word creativity derives from the Latin word "creo," which means to create, bring forth. Throughout history, creativity has been regarded as divine inspiration. In

Greek mythology, the Muses were the goddesses of song, music, and poetry. They became goddesses who inspired artists, philosophers, and intellectuals. The word inspiration originated from the Latin "inspiro," which means to breathe in or breathe life into, and it is what initiates the mental activity to new thoughts and ideas. Inspiration is often associated with motivation and creativity and is a contributing factor to our ability to create new things.

Creativity as a phenomenon and trait has been studied in different fields, from different perspectives. In the early stages of psychological research, creativity primarily involved unconscious spontaneous processes, operating under partial conscious control. There have been various explanations for the creative process. Creativity has been explained to be an associative and bisociative process, an interaction between different thought processes, a cognitive state characterized by heightened awareness, intense concentration, linked to a pleasurable activity, a sense of meaning and motivation, as well as a method of lifting mental blocks.

Sarnoff A. Mednick

The American psychology professor Sarnoff A. Mednick (1928-2015) explains creativity in terms of an associative process. In his book "*The Associative Basis of Creativity*" (1962), he describes creativity as a process in which associatively distant elements are connected into new combinations, creating new ideas. Creative individuals gather more associatively distant elements, and create useful associations between different ideas and concepts to a greater extent than less creative individuals. He developed a creativity test called the "*Remote Associates Test*" (RAT). It is a verbal test that measures the ability to make associations. The individual is given three words and is asked to come up with a fourth word that is associated with all three of them.

The Hungarian-British author Arthur Koestler (1905–1983) explains creativity from a different perspective than Mednick's theory. Kreativity is a process that involves a combination of elements drawn from two unrelated patterns of thought into a new pattern of thought, which he calls a bisociative process. In his book "*The Act of Creation*" (1964), he describes his theory of bisociation.

Sigmund Freud

Sigmund Freud (1856-1939) was an Austrian neurologist and the founder of psychoanalytic theory. Within the framework of psychoanalytic theory, two main forms of thought processes related to creativity are described. One is called primary process thinking, which is creative in the sense that it can generate new thoughts and images. It is a form of thinking typical of the unconscious mind, detached from being logical, rational, and realistic. It is characterized by being impulsive and jumping between different thoughts and associations. These are uncensored and uncontrolled thought processes driven by instincts and desires, aiming for immediate satisfaction of needs or wishes.

The other is called secondary process thinking, which is creative in the sense that it can create and refine new thoughts and images into a conscious rational form. It is a form of thinking typical of the conscious mind, connected to the principle of logical rationality and reality. It is a limited, controlled, censored and logical thought process that consider the demands for order, causality, and consequences.

Creative production involves interaction and cooperation between primary processes and secondary processes. It begins with primary process thinking, where unconscious thoughts and driving forces flow freely and generate new ideas and associations. Secondary process thinking can then process the creative thoughts. This involves applying logic, realistic thinking, and judgment to make thoughts useful. This step allows the individual to transform their creative impulses and ideas into something concrete and productive. The uncensored mind can, with increased age, collaborate with the disciplined, knowledgeable, and critically capable logical mind.

Mihaly Csikszentmihalyi

The Hungarian-American psychology professor Mihaly Csikszentmihalyi (1934-2021) links creativity to creative activity and the state of flow. He introduced the concept of flow after several years of studying artistic creativity and the personalities of artists. Flow is described as a state characterized by heightened awareness, intense concentration, and altered perception of time. Goals and strategies to achieve goals are logically connected. The activity proceeds smoothly and efficiently while there is control. Mental activity in all channels, which are not relevant to the current task, shuts down, making it possible to meet higher demands efficiently without becoming exhausted. The individual is concentrated on the creative activity, forgetting time. Awareness of oneself ceases, and thoughts about oneself become irrelevant, making the process much faster and more efficient. Flow is associated with a pleasurable activity, a sense of meaning, and motivation. In flow, the individual is engaged in the activity for its own sake. In the book "*Beyond Boredom and Anxiety*" (2000), the concept of flow is explained as helpful in understanding what makes activities motivating for humans.

Flow is not limited to certain areas. Flow can occur during any activity as long as its fundamental components are in place. There are several areas where the significance of flow has been noted. Examples of areas where interest in flow has been observed include sports, game-based studies, media usage, education, and science..

The Hungarian-Swedish microbiologist Georg Klein (1925-2016) has given a description of how authors and prominent researchers view the concepts of creativity and flow. According to Klein, flow is connected to professional skills and expertise. Flow leads to high motivation to tackle challenging tasks. It is characterized by intense concentration, accompanied by a euphoric feeling, and a sense of strong purpose. Scientific work and problem-solving proceed smoothly and efficiently while there is control. Scientific work and problem solving goes smoothly and efficiently, while there is control. He writes that the strong drive to take on a challenging task with one's own ability is a fundamental need of every human being. We want to experience the euphoria that flow can bring, and cannot experience the same euphoria by just repeating things. The challenge must be greater than that. We are driven by constantly learning new things. Although flow is an enjoyable and rewarding state, it involves a high degree of activation and thus needs to be balanced with relaxation and rest.

Brainstorming

In what is called "brainstorming", creativity is described as a cognitive state where mental blocks have been lifted, and a methods to overcome such blocks. It is a method aimed at solving problems or generating new ideas. Brainstorming is used in many areas, including product development and advertising, where the goal is to generate new ideas to solve problems. The method was introduced by the American advertising executive Alex Osborn (1888-1966). He wrote several books on the subject of creativity, and introduced the concept of brainstorming in the books "*How to Think Up*" (1942) and "*Your Creative Power*" (1948).

There are different variants of brainstorming, but all rest on the assumption that critical thinking blocks the emergence of ideas. Even though there are several variants, brainstorming is usually conducted by gathering a group of five to twenty people to work on a problem or task. The basic idea is that thoughts should run freely without being interrupted. Ideas should not be criticized before they are complete. The starting point is that all ideas are good ideas. By being spontaneous and non-critical, thoughts and ideas can run freely. An idea that initially does not seem good can develop into a brilliant one. All participants can develop and improve each other's ideas. The quantity of ideas is important because it is believed to generate quality.

Scientific studies have shown that brainstorming does not work as intended, including research conducted by the American psychologist Charlan Nemeth. One factor is that ideas are not allowed to be criticized before they are complete. Releasing critical thinking has been shown to result in fewer and lower-quality proposals. In meetings where ideas were allowed to be criticized, twice as many suggestions were made, and the following day, this group had seven times as many proposals.

In her book *"In Defense of Troublemakers: The Power of Dissent in Life and Business*" (2018), Nemeth explains why disagreements should not be feared but embraced. Nemeth argues that we make better decisions by embracing dissent because it forces us to consider more information and engages us in creative decision-making. It makes us question our assumptions and brings us closer to the truth, whether those assumptions are right or wrong.

Graham Wallas

In the late 19th century, researchers began identifying different stages or phases in scientific research. Graham Wallas (1858-1932) was a British psychologist known for the first model of a creative problem-solving process in scientific research. In his book "*The Art of Thought*" (1926), he describes four recurring stages or phases, which he calls preparation, incubation, illumination, and verification. Preparation is the phase in which the researcher deliberately gathers the information that needs attention and attempts to define the problem.

Incubation is the phase in which the information collected during the preparation phase flows freely and is processed unconsciously. There is no conscious focus on the problem; instead, attention is directed to another type of activity. The problem is set aside for a short or long period.

Illumination is the phase in which new insights emerge from the information processed during incubation. It is the moment when the information has been integrated, and the understanding of the problem's solution becomes clear. New insights cannot be calculated or thought out; they come like a sudden flash that illuminates the mind, hence the term illumination. An illumination can happen at any time and can never be planned. Verification is the phase when new insights are tested for validity and given form. It is the phase when the researcher determines whether the new insight leads to a solution to the problem and whether it can be further developed in detail. It requires conscious work, calculations, and discipline.

Many researchers have reported that thinking plays a subordinate role in the decisive moment that precedes the emergence of new insights. Creative insights come only after the researcher has stopped thinking about the problem, taken a break, or engaged in a less demanding task unrelated to the problem. It is important to take a break to gain a new perspective on the problem and avoid getting stuck in irrelevant details. Only when the researcher frees their attention from the problem can creative insights arise spontaneously and naturally. The brief moment that precedes a creative insight is called the creative pause. What happens during the creative pause has been the subject of debate. Some argue that creative insights originate from unconscious processing and integration of information. Others believe this explanation is insufficient and that a conscious space is needed for ideas to flow freely and fertilize each other.

Joy Paul Guilford

Joy Paul Guilford (1897-1987) was an American psychologist, known for his psychometric factor-analytic study of intelligence. In his book "*Intelligence, Creativity, and Their Educational Implications*" (1968), he presents his theory of intellectual processes in problem-solving. According to Guilford's Structure of Intellect (SI) theory (1955), an individual's performance on intelligence tests can be traced back to underlying mental abilities or factors of intelligence. SI theory comprises up to 180 different intellectual abilities organized along three dimensions: operations, content, and products.

SI includes six operations or general intellectual processes: cognition, memory recording, memory retention, divergent production, convegent production and evaluation. Cognition is the ability to understand, perceive, and become aware of information. Memory recording is the ability to encode information. Memory retention is the ability to recall information. Divergent production is the ability to generate multiple solutions to a problem; creativity. Convergent production is the ability to deduce a single solution to a problem; rule-following or problemsolving. Evaluation is the ability to judge whether information is accurate, consistent, or valid..

The distinction between convergent and divergent production remains a crucial idea in creativity research. Divergent production is a way to solve a given task by breaking logical thinking. It involves going in different directions and generating multiple solution ideas through rapid, intuitive processes. Tests measuring divergent production consist of tasks requiring the participant to come up with as many ways as possible to interpret or use an object of spatial, numerical, or verbal nature.

Convergent production represents the ability to find the correct way to solve a given task through logical thinking. It involves going in a single direction and generating a single solution through slow, analytical, and evaluative processes. Tests measuring convergent production consist of tasks requiring the participant to come up with the correct way to interpret or use an object of spatial, numerical, or verbal nature. Thus, a distinction arose between creativity and intelligence as two separate cognitive functions, with one associated with innovative thinking (creativity) and the other with logical thinking (intelligence).

Divergent and convergent production can be seen as two different cognitive styles in creative problem-solving. Individuals with a divergent cognitive style prefer to create something new, come up with new solutions to a problem. They are open to different perspectives, think of various ways to solve problems, and can see many possible approaches to a problem or task. Individuals with a convergent style prefer to analyze, evaluate, and find the right solution. They judge and weigh each possible idea to choose the best solution and use logical thinking to arrive at the correct solution to a problem.

Cognitive style refers to the preference or tendency to use certain mental strategies, methods, or patterns in processing information and solving problems. It is a concept used in psychology to describe people's different ways of thinking and handling cognitive tasks. Cognitive styles are neither good nor bad; they vary depending on the situation and task. They are unique ways of understanding and managing the world around us.

The concept of cognitive style was first proposed by the American psychologist Herman Witkin (1916-1979). In 1949, he published research results showing two distinct ways of perceiving phenomena in the environment, which he called field dependence and field independence. Some individuals are strongly influenced by the situation and context when they perceive and interpret something (field dependence). Others are more analytical and judge things without being influenced by the situation and context in which they occur (field independence).

Edward De Bono

Edward De Bono (1933-2021) was a Maltese-British physician, psychologist, and author known for his contributions to creative thinking and problem-solving. In his book "*Lateral Thinking*" (1979), he describes a model for two different approaches to problem-solving called vertical and lateral thinking. Vertical thinking is logical, language-based thinking based on preconceived notions and past experiences. It involves thinking within existing perceptions and conceptual frameworks, leading us into conventional and conditioned ways of thinking and perceiving. Perceptions and conceptual frameworks have an organizing effect on how we think and interpret the world. When we follow language-based logical thoughts, we are led into traditional thought patterns.

To break conventional thinking and generate unusual thoughts, a start from a point beyond the intellect's perceptual and conceptual limitations is required. Truly original ideas can only arise through a provocation that breaks language-based logical thought chains. Lateral thinking is a thinking technique that breaks language-based logical thought chains through provocation. By provoking, a new starting point is artificially created, and a new sidetrack is generated. While vertical thinking focuses on finding solutions to problems, lateral thinking focuses on finding new perspectives on problems.

According to De Bono, problems can be difficult to solve if we rely solely on vertical thinking because it limits our field of view to what we already know and is established. By introducing lateral thinking, we can open ourselves to new possibilities and find creative solutions that would not be apparent through vertical thinking alone. De Bono emphasizes the importance of being able to use both vertical and lateral thinking. Both approaches are valuable and can be used in combination depending on the situation and the type of problem that needs to be solved. Vertical thinking is useful in solving well-defined problems, while lateral thinking is useful when faced with complex and poorly defined problems requiring innovation.

In his book "*Six Thinking Hats*" (2016), De Bono provides a simple method for a leader to increase creative production and collaboration within a group and avoid long, unstructured meetings without making decisions. The method has been shown to reduce the length of meetings and decision-making time. This extremely simple method not only increases efficiency in discussions and decision-making but also results in a more pleasant working environment.

Research on creativity has documented that it is too simplistic to explain creative production with a single type of cognition. Creativity is a product, that does not come suddenly and unexpectedly, but is rather a consequence of thought processes, and involves an interplay between different mental processes. The creative process consists of two main types of mental processes, which are called generative processes and exploratory processes (Finke, Ward & Smith, 1992). It is a model that involves interplay between an initial generative phase, without strong goal orientation, followed by an exploratory phase with strong goal orientation.

Generative processes give rise to initial solutions, which are then repeatedly subjected to exploratory analysis to ensure solutions relative to the final goal. The generative process is fast, unconscious, and uncontrolled. It generates initial solutions, decisions, and judgments based on associative cues and pattern recognition. In the generative phase, many possible new solutions can be perceived. It cannot be interrupted or altered through rational thinking. The exploratory process is slow, conscious, and controllable. It determines the value of solutions, decisions, and judgments. In the exploratory phase, each solution is considered concerning the final goal. It can be interrupted or altered through rational thinking.

In the generative phase, mental representations are created, which are anticipatory structures. Generating anticipatory structures is only a small part of the creative process. Anticipatory structures must be interpreted and evaluated in the exploratory phase within the framework of the goal's constraints. In the exploratory phase, the anticipatory structures can either be focused on or built upon, and ideas can be analyzed, evaluated, and solutions can be constructed. In some cases, the analysis and evaluation will require a return to the generative process. Many leave the generative phase too early. It is a continuous cycle between generative and exploratory phases until a creative result is constructed. It may take several cycles before a final result in a creative product is achieved. The model presupposes an alternation between generative and exploratory phases and a refinement of structures in accordance with the requirements or constraints of the specific task.

Teresa Amabile

The American professor of business administration Teresa Amabile describes creativity in her book "*Creativity in Context*" (1996) as a method of problem-solving that relies on new adaptive responses to heuristic tasks. These responses can vary from a thought to being a product of artistic, political, academic, technical nature, or belonging to business.

A creative product is the result of three interacting functions. The first is having knowledge within a domain. The second is having knowledge of creative processes, which are methods to think innovatively. The third is motivation. When these three interact, the possibilities for a creative product are greatest. A product is considered creative if appropriate observers independently agree that it is creative. Appropriate observers are those who are familiar with the domain in which the product is created.

Amabile describes a frame for creativity consisting of three main elements: domain-relevant skills, which include actual knowledge, technical skills, and talents in the domain; creativity-relevant skills, which include cognitive styles, the application of heuristics, and work style; and task motivation, which determines the personal attitude towards a task. All three components operate at different levels in the process. According to Amabile, the most important component is the motivation for the task, which is not about external stimulus, but rather about an internal motivating state.

Creativity has often been associated with generating new ideas, but rarely with reference to the value and meaning of ideas. When a new idea is useful and feasible, it is called innovative, and when it is realized, it is called an innovation. An innovation is a product that has not previously been officially perceived and is significant for humans and society. This applies to all areas. It can be the manufacture of a product in a new material, using an object in a new way, or a theory that leads to the development of new technical aids.

Albert Rothenberg

Albert Rothenberg is an American professor of psychiatry who has focused on the creative process, emphasizing both novelty and value. He has conducted long-term studies of the creative process as it manifests in literature, art, and science, focusing on individuals who have been officially recognized and defined as creative.

In his research on scientific creativity, he focuses on a cognitive process he calls the Janusian process. It is a process characterized by consciously constructing and holding incompatible opposites and antitheses in the mind. The Janusian process is a type of thinking that should not be confused with dialectical thinking. Dialectical thinking is characterized by opposites coming serially, then being combined in a so-called "thesis-antithesis-synthesis." In a Janusian process, opposites are not combined but remain in unresolved conflict. They are held in tense opposition, operate side by side, and can generate new and valuable constructions in later stages. They can represent each other and be freely combined in various ways.

The Janusian process is purposeful and intentional. Concepts that bring together opposites or antitheses are consciously constructed. They are held in the mind throughout the creative process. The creative thinker is aware of conflicts among opposites that operate simultaneously and of contradictions from a strictly logical perspective. Opposites or antitheses develop through a complex series of mental processes and functions.

In the article "*The Process of Janusian Thinking in Scientific Creativity*" (1996), he describes his study based on interviews with twenty-two Nobel laureates in natural sciences. The study's results show that the Janusian process is a crucial ability during the creative process. Representations of opposites or antitheses are intentionally and consciously coexisting. Previously accepted laws, factors, and conditions are still recognized and judged as valid, but antithetical laws, factors, and conditions are formulated as equally functioning and valid. Although they seem illogical, the formulations are constructed logically and rationally to produce creative effects.

All participants were aware from an early stage of the significance and value of exploring the new. There was a strong motivation to create something new and a clear personal emotional engagement. This conscious motivation and intention to create is a characteristic of all creative individuals, regardless of the field they work in, says Rothenberg.

5. INTUITION AND RATIONAL THINKING

The word intuition derives from the Latin word "intuire," which means to gaze, observe, or see. Intuition is immediate insight or inspiration. Generally speaking, intuition is knowledge or cognition that is not preceded by logical, analytical, or rational reasoning. It is our ability to understand or draw conclusions without rational thinking. Intuition is a subjective knowledge and judgment that does not require reflection and analysis associated with rational thinking.

Intuition is a quick, unconscious process where we rely on feelings for judgments and decisions. It is part of the human psyche and is a key component in how we gather information, perceive, understand the world, and make decisions. It is a way of knowing something without being able to articulate it or know where the knowledge comes from. It is a feeling of knowing without direct sensory or perceptual content.

Intuition can be particularly useful for judgment and decision-making in situations with complex information, insufficient information, or when there is no time for logical analysis. Intuition has been used in computer science as a method in designing interfaces to facilitate smooth interaction between humans and computers. By understanding how people think and perceive intuitively, interfaces can be tailored for humans.

In knowledge development, intuition has been used as a method in perceiving patterns and interpreting meaning in texts with complex information. Intuition has been used as a method in scientific knowledge development in the creation of new research areas and as an initial step in an as-yet undefined research project.

There are various theories about intuition. Intuition has been explained as a global structural cognition based on the complex selection and organization of information. Intuition has a unifying effect on rational reasoning when it comes to choosing hypotheses and problem-solving, and it provides a sense of certainty to rational reasoning. Intuition has been explained as both separate from and related to intellectual analysis and logic. Two philosophers with opposing views on intuition in relation to intellectual analysis and logic are Henri Bergson and Hans Larsson.

Henri Bergson

Henri Bergson (1859-1941) was a French professor of philosophy and a Nobel Prize winner in literature in 1927. According to Bergson, intuition is a more fundamental perception of reality than the intellect. It is not possible to reduce intuition to abstract concepts and intellectual analyses. The intellect only distorts our perception of what is real, while intuition gives us the ability to experience reality directly. Intuition is a spontaneous, immediate perception of existence as a whole without the involvement of intellectual activity. It is a way to intuitively understand, allowing a deeper insight into existence. By the act of intuition we feel sympathy with every states of the object and this sympathy paves the way for absolute knowledge. In short, according to Bergson, intuition is simple experience of sympathy, namaly going into an object and grasping its uniqueness and peculiarity.

Intuition is central to human understanding and knowledge and should be a source of inspiration for both philosophers and physicists. Traditional philosophy and science limit our understanding and knowledge. Bergson criticizes their goals and methods as dealing with a dialectic of words based on conceptual categorizations and analytical thinking. To be a genuine search for truth, philosophy and science need to be rooted in experience.

Hans Larsson

Hans Larsson (1862–1944) was a Swedish professor of philosophy, known as a cultural debater and educator. According to Larsson, intuition depends on intellectual analysis and logic. Intuition can only appear independent of intellectual analysis and logic, but this is only how it seems. Intuition is perception of complex relationships, a finer and more nuanced form of logic. Intuition is the ability to understand a wealth of representations in a single perception and is the highest cognitive function. This is only possible if knowledge and experience have been sufficiently processed so that thought and feeling can unite. Intuition is not only a way to knowledge but also a way to live a more authentic and meaningful life. In his philosophical works, he often refers to literature, in which he finds examples of this high-level intuition.

Carl Gustav Jung

The Swiss psychiatrist Carl Gustav Jung (1875-1961) is known for his theory of different personality types. Jung divides the personality into four basic functions, which he calls thinking, feeling, sensing, and intuition. These are related to how people gather information, perceive the world, and make decisions. There is always some combination of these types that dominate the human psyche. Thinking and feeling manage our decision functions. They are rational in the sense that they guide us in what is right to do. For a thinking person, the rational means what is logically right, while a feeling type experiences the significance of what feels right.

Sensing and intuition manage our perceptual functions. They are irrational in the sense that they relate to our perception. Sensing provides sensory impressions and the perception of objects, while intuition allows us to directly perceive something and offers a more general understanding of contexts without having facts or specific data as a basis. If sensing means perceiving through our senses, intuition means feeling potential of something.

Jung emphasizes that people have different preferences and strengths regarding these functions. Each individual has a dominant function, meaning they have a natural preference for using one of the four function types when thinking and acting. One of these functions will become the primary conscious function, while one of the others becomes the secondary. An individual can only handle a couple of the functions consciously. Some individuals may be more inclined to use intuition as their primary function, while others may prefer thinking or a combination of the two.

For example, an individual may have a preference and aptitude for thinking and intuition. Thinking can help systematize ideas, while intuition can help determine the purpose of these ideas. Intuition is a source of creativity and inspiration and a significant part of the human psyche's experience of understanding the world. Through intuition, we have the ability to perceive potential possibilities and implications of situations, objects, and actions, and it initiates ideas, symbols, and analogies in creative problem-solving. Intuition can be suppressed by logical constraints, but we need logical thinking to organize, communicate, and validate intuition.

Rational Thinking

The word "rational" derives from the Latin word "rationalis," which means "reasonable." To be rational means to use common sense, reason logically to make decisions, and draw conclusions based on facts and analysis. It is the conscious thought process that an intelligent being uses to solve problems and achieve goals. It is a deliberate, intentional, and goal-directed thinking involving the intellect, knowledge, and experience. Rational thinking and decision-making are the highest intellectual functions. It is a complex function where decisions are made by considering various solution alternatives, weighing the pros and cons, and making decisions in line with the desired goal. The goal being pursued determines how the problem is defined. Often it is more challenging to identify what the problem is than to find solutions to the problem.

Rational thinking works well for solving problems and achieving goals when the conditions assumed are known at the time of implementation. The conditions applied when the solution is determined may have changed, and new conditions may have emerged. It is not always possible to control and foresee new conditions. New facts and other goals may have emerged at the time of the solution's implementation. The fact that a solution works well in one context does not necessarily mean it will work well in another context. Rational thinking, like intuition, is not a guarantee that something is true or likely. It can often be used to justify our beliefs and belief systems. We can reason logically based on our beliefs and declare them true.

It is not enough to claim that something is true. Reasons and counterarguments are also needed to determine if a statement is true. Providing reasons or counterarguments is called argumentation. Argumentation means to defend a statement or opinion based on certain assumptions (premises). The premises on which the argumentation is based must be relevant to the argument's conclusion. Presenting an argument based on an unknown or unproven premise is called a fallacy or error in reasoning. This means that the conclusion drawn does not logically follow from the premises presented.

There are different types of fallacies. Fallacies can take the form of circular reasoning. This means that reasoning uses its own statement to prove its validity and uses arguments starting from what needs to be proven. Another fallacy is assuming a causal relationship between events when they occur sequentially. That something happens after something else does not necessarily mean the preceding event caused the subsequent event, but the connection may depend on underlying factors.

Another common fallacy is the assumption that if a theory is correct, it implies that all other theories are incorrect. No theory needs to be correct, or there may be other theories that are also correct. Other types of fallacies may involve letting the credibility of a statement be determined by who argues for or against the statement. This means discussing the opponent as a person instead of the current subject, as it is easier to focus on a person's characteristics or opinions to dismiss arguments. It is to use emotional arguments and aversion based on fear of the opponent's message. It is always possible to identify fallacies in one's thinking and train oneself to think more logically. Like most skills, the ability to think rationally and logically can be developed and improved.

Daniel Kahneman

Daniel Kahneman is an Israeli-American professor of psychology and a Nobel laureate in economics (2002). He is considered one of the founders of the field of behavioral economics. In his book "*Thinking, Fast and Slow*" (2011), Kahneman describes two distinct cognitive systems for thinking and their roles in decision-making, which he calls the fast system (System 1) and the slow system (System 2). The two systems can interact, block each other, or alternate depending on the situation and needs. Both have their weaknesses and often conflict with each other.

System 1 is fast, intuitive, and emotional. It involves unconscious and uncontrollable cognitive operations. It generates automatic interpretations of events, provides quick explanations for what is expected, and seeks causes for what is unexpected. It operates without conscious deliberation and effort. This system enables immediate associations, recognizes faces, interprets simple patterns, and makes decisions based on experience and intuition.

Intuition is characterized by the cognitive ease with which it processes information but does not signal when it becomes unreliable. Answers come quickly and confidently but offer no assurance of being right or wrong. Intuition can lead to many correct conclusions but also many mistakes. Not everything that feels convincing and right is necessarily true. We overestimate certainty and overlook the facts needed for decisions. Intuition can provide a quick, confident response to situations, but overlooks the consequences relative to the desired goal.

System 2 is slow, deliberate, and logical. It involves conscious and controllable thought operations. It is a system that works with abstract concepts and solves complex tasks. It is used for solving complex problems, performing mathematical calculations, and making decisions based on logic and facts. It is characterized by slow, attention-demanding processing of information. It involves carefully considering information, weighing different options, and using logic and evidence for decision-making. To improve performance in activities such as judgments and choices, attention investment, concentration and critical thinking are required.

The ability to think rationally is not a guarantee against misjudgments. Deficiencies in rational thinking can stem from lack of concentration, insufficient knowledge, or superficiality and laziness in thinking. We can deviate from the rules of logical thinking in many ways. We do not always think strictly logically when reasoning and often make mistakes because we do not know better.

According to Kahneman, there is no simple way to resolve the conflict between intuition and rational thinking. Often, we seek rational reinforcement for our ideas and feelings after forming our intuitions. We look for facts or rational arguments that confirm our intuition because it is the easiest way out of the dilemma. Attempting to reconstruct rational reasons for our beliefs can easily lead to mistakes, but we can slow down and seek a rational explanation instead of trying to construct a rational answer. The best strategy is to use both thinking systems complementarily.

It is important to be aware of the inaccuracies that can arise in intuitive judgments. We are often inclined to use quick and intuitive judgments because they are less demanding and more efficient in our daily situations. If we are not aware of these risks, it leads to cognitive misinterpretations and erroneous decisions. By being aware and considering alternative perspectives, we can use both thinking systems in a way that leads to more well-founded decisions.

Hans Ole Rosing

Hans Ole Rosing (1944-2018) was a Finnish lecturer in the philosophy of science known for his contributions to exploring and explaining the logical foundations of scientific thinking. In his book "*The Logical Foundations of Science*" (1978), he addresses various aspects of the scientific method and logic. According to Rosing, logic in its broadest sense can be defined as any form of reasoning that leads to a conclusion. All forms of reasoning are ultimately based on assumptions we take for granted. We perceive them as true without deriving them from rational reasoning. They can be false and often not even conscious.

There are many different ways to reason, but not all are equally good. If all forms of reasoning are based on arbitrary assumptions, knowledge becomes relative. Each culture or belief system would then have its own logic. The most radical view is that each individual has the right to their own logic, reasoning based on their own assumptions. If every logical conclusion is correct in its own way, regardless of its assumptions, then all research becomes meaningless. Scientific research is meaningful only if we can be mistaken and wrong. If everyone is right in their way, there is no reason to discuss subjects we disagree on, and no objective truth we can seek. There would only be different opinions, all of which are right. When seeking factual knowledge and explanations, the scientific method of reasoning is the best, but that does not mean it is infallible. Not all problems can be solved with scientific logic. There is an objective logic, but some problems cannot be solved with it.

For everyday situations, there is only one correct logic, called intuitive logic, says Rosing. It is a type of logic we use when drawing conclusions without consciously applying formal logical rules. Intuitive logic is based on humans' natural ability to understand and reason about relationships and statements. We use it to make judgments and draw conclusions based on experience, background, and knowledge. It is based on linguistic statements that most of us immediately accept as true. What is true is a consequence of the meaning of words and how we use them. We understand the meaning of words without basing them on anything else.

Intuitive logic is the same type of intuition as we have about our native language or simple mathematics. and is a prerequisite for knowledge based on logical thinking to be possible. This ability can be trained through practice, just like the ability to speak and write one's native language. All logic is ultimately based on intuitive logic and a "feeling" for what is right and wrong.

If intuitive logic did not exist, logical thinking would not exist either. The ability to draw logical conclusions has been an effective tool for biological survival. Among our ancestors, this ability developed as an effective weapon in the struggle for existence. A people without this ability would hardly survive for long. Even animals and birds display simple behavior based on logical inferences.

Intuitive logic differs from formal logic in handling of logical arguments and relationships. They have different methods for reasoning and drawing conclusions. Intuitive logic is a more informal and everyday form of logical thinking, while formal logic is a more structured and precise method for studying and analyzing logical arguments. Both forms have their uses and values, depending on the context and purpose of the reasoning.

All science is based on intuitive logic. The relationship between intuitive logic and the logic of science can be compared to the relationship between theoretical mathematics and practical everyday mathematics. We do not need theoretical logic to reason logically about everyday life problems, just as we do not need to know algebra to calculate our income and expenses. All theoretical reasoning must be based on intuitive logic. Our intuitive sense of what is right and wrong is the foundation we always ultimately rely on. If we step outside this foundation, there is no difference between right and wrong, true and false, meaningful and meaningless.

Gottlob Frege

Gottlob Frege (1848-1925) was a German mathematician, logician, and philosopher, known as one of the founders of modern formal logic. Frege has had a significant influence on mathematics and philosophy in the early 20th century and is considered one of the most influential thinkers in the history of logic and philosophy. Frege's work in logic and philosophy has been groundbreaking and is still considered relevant in philosophy and logic. His ideas have profoundly impacted developing modern formal logic systems, and he is regarded as the first to introduce formal logic. During the 20th century, formal logic developed rapidly, and many new formal systems were created. Nowadays, there are many different formal systems and applications of formal logic, one of its most important applications being in computer science.

Formal logic is a more structured and abstract form of logic. It uses carefully defined rules and symbols to analyze and evaluate logical arguments. Formal logic studies logical arguments precisely and determines whether a conclusion is valid or invalid based on its premises. It handles constructed symbolic languages and logical inferences independent of the content of linguistic entities and resembles mathematics, to which it is closely related. A set of axioms is formulated in a symbolic language, and with specific rules for derivation, formulas can be derived.

Even formal logic is based on intuitive logic. Creating formal systems without connection to intuitive logic would be like writing texts by putting words in a row according to an invented grammar not based on any natural grammar. It is important to distinguish between everyday logical reasoning and formal logic. Formal logic is a simplified and idealized form of our everyday logical reasoning. Everyday logical reasoning is never formal. Instead, we use natural language when we are drawing conclusions. It can be formalized into a formal system, but it does not make us better at thinking logically.

6. COGNITIVE SCIENCE AND DEVELOPMENT

Research on thinking is divided into two main areas cognitiv science (thinking as a computational process) and cognitive development (thinking as intellectual development). Cognitive science studies the nature of human thinking, primarily focusing on how people receive, process, and use information. It involves problem-solving and decision-making by organizing, storing, retaining, and recalling information. The starting point is how individual associations and experiences affect the information process. The focus is on cognitive processes such as memory, perception, attention, problem-solving, and decision-making. One goal is to construct algorithms that can replace problem-solving, rational thinking, and decision-making. This is based on the assumption that thinking and decision-making can be simulated through the use of mathematical or logical symbolic languages.

Initially, cognitive science was influenced by the development of computer science. The computer was used as a metaphor for cognition, understanding thinking as data processing. According to the traditional cognitive science model, cognition is equated with information processing in a computer, and computers were used as tools for cognitive modeling. By constructing models that can be executed on a computer, various types of theories could be tested.

Modern cognitive science has moved away from the strong influence of computer science and the computer metaphor. The concrete situation is considered important for cognitive assessment and and human reactions. This perspective is called situated cognition. It means that it is by processing impressions from specific details in the situation and making a judgment based on experience that actions are determined. The concept of situated cognition has led many to question the necessity of mental representations to explain cognitive behavior. More emphasis is now placed on the mutual influence between the brain, body, and environment rather than cognition as passive information processing. Cognitive science does not include aspects such as context-dependent factors, historical factors, and emotions. It has been questioned whether it is possible to isolate thinking from emotions and motivations. Research on how people process, react to, and use information cannot lead to an understanding of all aspects of the psyche.

Cognitive Development

Cognitive development studies the development of thinking. It involves the gradual and systematic change in an individual's ability to think, understand, and solve problems with increasing complexity and capacity over time. This perspective is central to understanding how individuals acquire and use knowledge and how they interact with and adapt to their environment.

Based on experimental studies and observations, different theories about the maturation and development of thinking emerged in the 20th century. The research was mainly based on cognitive development during childhood and adolescence, primarily the development of cognitive abilities related to how the individual thinks, understands, and functions intellectually. Cognitive development was divided into different age periods, linked to the brain's neurological development, and attempts were made to find consistency between these age periods and the brain's neurological development.

There are different opinions about what these age periods are and when intellectual abilities emerge. Understanding cognitive development became central in education and pedagogy as it was believed to help teachers and educators adapt their teaching methods to the learners' age and abilities. Cognitive development became an important concept in psychology and pedagogy and has been studied by researchers such as Jean Piaget and Lev Vygotsky, both known for their significant work in developmental psychology, and remain influential in education and pedagogy.

Jean Piaget

Jean Piaget (1896-1980) was a Swiss mathematician and psychologist, known for his intellectual perspective on cognitive development. He is considered a pioneer in developmental psychology, and Piaget´s theories have influenced our understanding of how human intellect develops. His theory of cognitive development is based on the idea that children go through different stages of intellectual development as they grow and learn. Each stage has its characteristic way of thinking and perceiving the world, forming the basis for the development of the next stage. Piaget identified four main stages: the sensorimotor, preoperational, concrete operational, and formal operational stages.

The sensorimotor stage occurs from birth to about two years. During this period, the child explores the world through their senses and motor skills. The child develops perceptions created from sensory impressions and motor skills. By using their senses and motor skills, the child seeks to create order from what they perceive as chaotic.

The first thought structure allows the child to coordinate sensory impressions and movements. The child develops object permanence and the invariance in the relationship between a specific movement and a specific effect. With the ability to perceive permanence, the child gain stability in their environment.

The preoperational stage lasts from about two to seven years. During this period, the child develops the ability for symbolic thinking and imagination. Language develops along with the ability to handle concepts such as children, parents, play, toys, and more. The ability for internal representation of external objects and events develops, as does the ability for self-awareness and other representations. Thinking is still egocentric and lacks conservation (the understanding that the quantity of a substance does not change even if its form changes).

The concrete operational stage occurs from about seven to eleven years. During this period, children become better at concrete logical thinking. They understand the concept of conservation and can perform simple math. The child develops the ability to think logically in a concrete way. They can understand logical and mathematical relationships between weight and quantity, density and weight, length and width, but only when the objects are presented concretely. An understanding of other people's ways of thinking also begins to emerge.

The formal operational stage occurs from about eleven years and onwards. During this period, individuals can think abstractly, solve complex problems, and use logic and reasoning to handle theoretical questions. Understanding develops for abstract logical reasoning and abstract concepts. The ability to reflect on one's own thinking develops, as does the ability to plan and think ahead. The individual can perceive the logic in abstract thought operations and understand logical and mathematical relationships without needing to observe objects physically. Abstract logical thinking was the highest form of thinking.

According to Piaget, the individual attempts to achieve a coherent understanding of themselves and the world through their mental structures. It is the individual's need to understand and find meaning in the world that drives the development of thinking. The theoretical starting point is that there are universal forms of knowledge structures that govern an individual's understanding of the world. Understanding the world is linked to individual thinking, and the individual learns through interaction with the external world.

Piaget was fundamentally a biologist and developed his cognitive theories based on the idea of biologically open systems. Just as such systems develop, so does an individual's intellectual development through interaction with the external environment. Cognitive developmental stages are biologically determined and associated with different age periods linked to the brain's neurological development. It means that children cannot assimilate knowledge that is too advanced for their biological development.

Piaget showed through his studies that children's thinking and adult thinking are fundamentally different. Children's way of thinking and intellectual functioning develop gradually, not through a gradual accumulation of experiences. He deviated from the prevailing view that children's thinking was a less developed version of adult thinking, especially concerning developmental steps related to logical thinking. Children construct knowledge through interactions with the external world based on their maturity level to handle these interactions. Their knowledge is constructed through a motivation that comes from within them. Piaget believed it was important to let children explore and discover knowledge in their own way, in line with their cognitive developmental level. This has influenced pedagogy by adapting educational methods and materiels to children's developmental levels.

Piaget's theory has been criticized for focusing solely on intellectual development and abstract logical thinking. By only focusing on the development of thinking, he is considered to overlook other factors that can influence development, such as environmental and cultural factors.

Lev Vygotsky

Lev Vygotsky (1896-1934) was a Russian psychologist and educator known for his sociocultural perspective on cognitive development. Vygotsky argues that children's brains do not only develop for abstract thinking. The surrounding culture and the tools offered also influence the development of thinking. It is more about interaction with the surrounding environment than an independent individual process. Social factors, such as interaction between the child, peers, teachers, parents, and the surrounding culture, also affect development, not just biological factors. The focus is on how social interactions, cultural factors, and social contexts influence cognitive development, how we think and learn. Children learn through interaction with other people, including adults and peers. Social interaction is a key component of cognitive development.

Vygotsky provides a cultural-historical or sociocultural perspective on learning, as he sees the environment as crucial for an individual's development and performance. Cognitive development occurs through relationships with others, especially parents, and a striving for cultural understanding (conceptual understanding). Mental activities begin with social interactions and then become internalized as independent mental activities. There is a difference between what children can learn on their own and what they can learn together with a more competent person who can broaden their opportunities to learn. It can be a teacher, a peer, a parent, or anyone who can support children in handling tasks more independently. For children to develop maximally, they should be given tasks that are a little more logically complicated than what they have previously succeeded with.

He explored the relationship between scientific and everyday concepts. Scientific concepts are theoretical, while everyday concepts are based on experience. He shows how teaching involves intersecting these concepts, where the learners' experiences confront scientific concepts. It is in a complex situation in which the teacher's scientific concepts set the spontaneous thinking in motion towards a qualitatively higher level.

Vygotsky examined children's language development and has had a significant influence on research on children's language development. Language and thinking are closely connected with childrens development. There is an elementary pre-linguistic thinking in children, which is purposeful thinking. Children can mentally think out tools and similar objects before the development of language and speech. Signs of such purposeful thinking can also be observed in animals. Even animals can find new ways to invent and manufacture tools that solve their problems. In his book "*Thought and Language*"(1934), he describes his theory of the coherent development of thinking and language.

Vygotsky's theory of children's language development is based on the assumption that children learn through social, linguistic, and cultural interactions. Children need their social interactions to build language processes, and through these language processes, they develop mental tools to learn. Children learn by working with the physical and mental tools available in their culture. Social norms and culture influence what they will learn and how they will develop. Adults share their cultural norms and perceptions of the world with children. There are different tools for physical work and different mental tools. Different cultures use different symbols, languages, and tools to acquire knowledge. Cognitive development occurs when language is assimilated and becomes part of the child's identity. In cognitive psychology, language is an essential requirement for thinking.

Vygotsky has played a central role in developing modern pedagogy and psychology. His work has significantly impacted areas such as education and developmental psychology. He emphasized the importance of understanding learning and cognitive development as a social and cultural process and shaped how we think about teaching and pedagogy. His ideas remain influential in education and research.

Noam Chomsky

Noam Chomsky is an American professor emeritus in linguistics, often regarded as the father of modern linguistics, known for his groundbreaking work in the science of human language and generative grammar. Generative grammar is a collective term for syntactic theories in linguistics, aiming to describe and explain language. Syntax or sentence theory is part of grammar. It refers to how we put words together into groups of words, clauses and sentences. Chomsky's work has been highly influential, particularly in understanding how people think and acquire knowledge, and his theory of language has influenced disciplines such as psychology and philosophy.

In his doctoral dissertation from 1955 and later in the more well-known "*Syntactic Structures*" (1957), he argues that there is an innate, universal grammatical structure common to all human languages. People share certain fundamental linguistic structures, regardless of the language learned. He introduces the theory that parts of human grammar are already inherited in the brain. We use innate structures of initial assumptions acquired from language, and based on these structures, the grammar of the linguistic community is constructed. It is the linguistic environment that determines which languages children acquire, but all children learn languages similarly, through universal language development. Children learn their mother tongue during the first year of life without being exposed to all possible structures of the language.

Humans' innate language ability and internal grammar form the basis of what is known in generative grammar as universal grammar. Since all humans quickly become proficient in understanding and using their mother tongue, all linguistic competence is biologically based, and thus there must be a universal grammar, a sort of innate blueprint of initial assumptions that form the basis for grammatical construction in our linguistic community.

His focus is on representing the underlying grammatical structure of natural language by accurately predicting which syntactic structures that can be possible and impossible in a given language. All language models must be able to predict which combinations of words can be considered acceptable and unacceptable to an adult native speaker.

Chomsky distinguishes between an individual's linguistic performance, which are the actual utterances in the language, and an individual's linguistic competence, which are the possible utterances in the language. Linguistic competence cannot be understood merely by describing the structure of the utterances we produce; rather, we must consider thematic relations described at a more abstract depth.

An important claim by Chomsky concerns the modularity of the mind. According to Chomsky, the mind is a set of autonomous processing units, each with its own set of rules specialized for domain-specific information. Information processing in each unit is encapsulated so that the unit's function is independent of activities in other units. In terms of language, this means that language processing occurs independently of all other intellectual activities. Regarding creative processing, there are units specialized and separated from language. Creative activity has its own units, each with its own mental operations. This means that creative activity is more like general intelligence. Many of his perspectives on the nature of the human mind have relevance to human creativity.

Chomsky's theory of generative grammar has undergone many changes, but has maintained its emphasis on grammar and everyday linguistic creativity. His work in generative linguistics contributed to the decline of behaviorism and the promotion of cognitive sciences.

He challenged the behaviorists' view of language development and argued that the human mind was more complex than behaviorism suggested. His criticism of behaviorism laid the foundation for a clear paradigm shift in the 1950s, involving a mentalistic approach to both language and cognitive psychology.

Cognitive Psychology

Cognitive psychology emerged as a reaction against behaviorism, which had been dominant in psychology during the 20th century. Behaviorism focused on what was measurable with studies directed towards observations of external behavior. Mental processes, such as thoughts and feelings, were not observable and where therefore considered impossible to study. For behaviorists, what happened in the mind and brain was of no interest. The brain was a passive organ reacting to stimuli from the external world. Cognitive psychologists, in contrast to behaviorists, argued that it was possible to explore even non-observable behaviors scientifically.

For cognitive psychologists what happened in the mind and brain was the most interesting. The brain was not a passive organ, but an active one reacting to stimuli from the external world. It was considered important to study what happens when stimuli reach our sensory organs, and which stimuli are attended to. Cognitive psychology focuses on the mental structures and processes we use to think and interpret the world. The focus is on how our intellect functions, influences behavior, and how sensory impressions from the external world cause changes in the mind. An important premise in cognitive psychology is that our way of thinking determines our subsequent reactions. How we react to the external world depends on thinking rather than external circumstances. Through our thinking, we create the reality to which we respond and react.

7. THE TWO BRAIN HEMISPHERES

The functional asymmetry between the two brain hemispheres has long been a debated topic. There are differences between the two hemispheres, but opinions on why they differ have varied. Speculations range from the idea that the two hemispheres are merely reserves for each other to the notion that they complement each other.

The English physician Arthur Ladbroke Wigan (1785-1847) was one of the first to note that humans have a dual brain and proposed a theory of "The Duality of Mind." When Wigan examined the brain of a man, he discovered that the man had only one half of a brain. Despite this, the man was able to speak, read, write, and function normally. Wigan therefore concluded that since the man had only half a brain and functioned mentally, it must mean that people with two brain hemispheres have two minds.

Roger Wolcott Sperry

Roger Wolcott Sperry (1913-1994) was an American neuropsychologist and neurobiologist who first conducted a scientific study of the brain's two hemispheres. He was awarded a Nobel Prize in Medicine in 1981 for his groundbreaking discovery of the distinct specialization of the two hemispheres. During the 1960s and 1970s, Sperry, together with the American neuropsychologist Michael Gazzaniga, conducted a series of experiments known as the "split-brain experiments."

These experiments were performed on patients who had undergone an operation known as corpus callosum sectioning, which means the nerve fibers that normally connect the two hemispheres had been cut. This operation was performed on patients suffering from severe epilepsy to prevent an electrical seizure starting in one hemisphere from spreading to the other.This meant that the two brain hemispheres could no longer communicate with each other. By studying how these patients reacted to different stimuli and tasks, they were able to show that the two hemispheres had different functions and could operate relatively independently of each other. The hemispheres had specialized functions, and information in one hemisphere could be processed without the other being aware of it.

Normally, the left brain hemisphere attends to information from the right eye and controls the right side of the body, while the right hemisphere attends to information from the left eye and controls the left side of the body. By sending visual information to only one hemisphere's visual field, their different ways of being conscious could be explored. In experiments, patients looked at a screen so that images or words displayed could only be seen by one eye, ensuring the information went only to the opposite hemisphere.

When visual information reached the visual field of the left brain hemisphere, patients responded by telling a story about what they saw. The left hemisphere, which housed a language center and speech capability, acted as an interpreter and tried to find an explanation for what it saw. When objects were shown to the left eye (and thus went to the right hemisphere), they could not verbally report the objects and words shown, instead responding by just pointing to an object.

Results from these experiments confirmed that the left brain hemisphere is involved in language comprehension and speech, while the right brain hemisphere is involved in visual and spatial processing. This indicates that the two hemispheres have different thoughts and perceptions of what they see and have their own notions of what is right. Several years of repeated experiments have provided evidence that the left and right hemisphere can be conscious in different ways. They process information differently and give rise to different ways of thinking and perceiving.

Not only do they think and perceive differently, but they also exhibit two distinct characters. Both the way they process information and intentions are different. They have different beliefs about what is right and exhibit different wills, that often are in conflict.

The result interpretes as that the two hemispheres have specialized functions. The left hemisphere is involved in linguistic processing and speech, while the right hemisphere is focused on visual and spatial processing. This impacts our understanding of how the brain functions and handles information. It also provides insight into how consciousness and perception can vary when the two hemispheres cannot communicate with each other. The experiments have significantly impacted our understanding of the brain's organization and function. One common adoption is that the left brain hemisphere has to do with logical, analytical verbal thinking, while the right brain hemisphere has to do with creative, intuitive, nonverbal thinking in images.

Oliver Sacks

The British-American neurologist Oliver Sacks (1933-2015) is renowned for his books on neurological case studies. He wrote about patients with neurological disorders in a manner that combined profound medical insights with compassion and humanism. Sacks helped popularize complex medical topics and make them accessible to a broader audience. In his book "*The Man Who Mistook His Wife for a Hat*" (1988), he provides an example of a case where a patient, due to damage in certain visual centers, could see various things without recognizing them and lacked awareness of this defect. The man was a prominent musician and a teacher at a music school. It was during interactions with his students that some of his strange problems first became apparent. Students could approach him without him recognizing their faces; only when they spoke could he recognize them.

These incidents became more frequent and caused great confusion. The man's difficulty in recognizing faces increased, and he began to see faces even where there were none. Initially, these mistakes were dismissed as amusing, especially by the man himself. He did not feel ill and thought the mistakes were too comical to be taken seriously. When Sacks examined him, he found no signs of deterioration. The man expressed himself freely and clearly, with imagination and humor, but then something strange happened. As the man was about to leave Sacks' office and knew he was supposed to put on his hat, he saw his wife's head and tried to lift it off to put it on his own. This was an example of the bizarre symptoms he exhibited when misinterpreting his visual impressions due to his brain injury.

Sacks questioned how this inability to interpret and judge should be understood and realized that it must be the concrete reality the man had lost. This seeing had been lost, even though his judgment was quick and normal in all other respects. It was his ability to see how things were related to himself and to each other that had been lost. His abstract approach to reality had prevented him from making an intuitive, personal, and concrete judgment. Abstract thinking is not the only path to reality and truth, says Sacks. The intuitive and personal is also a way to reality and truth. In contrast to the abstract, the concrete is also real but in a completely different way.

In neurology and neuropsychology, the concept of "judgment" is almost never discussed, even though the disintegration of judgment is central to many neuropsychological disorders. Our ability to judge is the primary faculty of higher consciousness, yet it is ignored or misunderstood in classical, data-oriented neurology. Our mental processes, which constitute our being and life, are not solely abstract; they are also personal and involve judgment and feeling, not just classification and categorization. A science that shuns judgment, the distinctive and the personal, becomes entirely abstract and data-oriented, leading to the reduction of the concrete and real.

In contrast to natural science, Sacks devoted his interest to the individual human being. We can navigate our life situations and survive thanks to our nervous system. By having language, we can transfer information from one generation to the next and from one continent to another, creating cultures, art, and science. All this is possible because of our ability to handle abstract symbols. Damage to the nervous system, small or large, localized or diffuse, changes both the structure and function of our brains. If our intellect and emotional life diminish, our personality changes, and we are no longer ourselves. Our senses constitute the sensory world, but there are also hidden senses that are just as important. These are unconscious, automatic senses that need to be discovered. If our perception of our hidden senses is damaged or disturbed, we experience something akin to being blind or deaf. The body becomes blind and deaf to itself, and we no longer "own" a self and don´t feel a self.

Each of us has a life story, an inner narrative whose continuity and meaning constitute our life. We construct and live in a story, and this story is our identity. Every individual has a biography, a history, and a narrative that is constantly and unconsciously built by, through, and within us through our sensory perceptions, feelings, thoughts, and actions, not least our verbal expression of speech. Biologically and physiologically, we do not differ much from each other, but as stories, each person is unique. No physiological or neurological assessments are needed to tell our story.

According to Sacks, knowledge of the functions of the two brain hemispheres has primarily been obtained through studies of damaged brains, particularly regarding damage to the left hemisphere. Neurology and neuropsychology can be seen as a whole history of the examination of the left hemisphere. There are attempts to explore the right brain hemisphere's syndromes, but strangely enough, these attempts have been overlooked. The right hemisphere has often been considered more "primitive" than the left hemisphere, which has been seen as the pinnacle of human evolution. The right hemisphere controls our ability to recognize reality, a capability that every living creature must have to survive. The left hemisphere is designed for programs and schemes. Neurology has been more interested in schematization than reality, which has led to that the right hemisphere's syndromes being considered bizarre when they manifest. Studies of the role the right hemisphere plays in direct consciousness have been neglected.

Jill Bolte Taylor

Jill Bolte Taylor is an American neuroscientist, that had a blood clot (a stroke) in her left brain hemisfere. In her book "*My Stroke of Insight: A Brain Scientist's Personal*

Journey" (2006), she describes her experiences both during and after her stroke, and the insigts of the brain and consciousness, she got by observing her condition efter the stroke. Her experiences gave her a deeper understanding of how the right hemisphere focus more on direct consciousness. Taylors experiences have provided insights into brain function and understanding of what it is like to recover from neurological damage. Taylor has, with a background in neuroscience, been able to analyze her experiences with a unique perspective. In the book she discusses the contrast between the two hemispheres.

When Taylor had the stroke, she woke up to the gradual loss of function in her left hemisphere. Over the course of four hours, she experienced her own brain breaking down. This led to her losing the ability to walk, speak, write, read, and remember anything from her own life, but she remained conscious and experienced the world through her right brain hemisphere. When she woke up the first morning, she remained conscious, but she no longer had the variety of skills that would normally be active in the left hemisphere. Moments of clarity occurred when she could reach a phone and ask for help. When she tried to call for help, she could not understand what was being said on the phone and could not express herself, even though she knew exactly what she wanted to say. She could not put words together into anything meaningful but could listen to the intonation, so she understood that she would get help.

Due to the damage in the left hemisphere's language center, the constant monologue in her consciousness fell silent. The inner voice was quiet for the first time, and she heard no voices telling her what to think, do, or who she was. There were no thoughts about the past or future and no goals to achieve. Memories dissolved, and she had no pain from the past or fear of the future, and no memory of previous relationships. She lived without being aware of time in a constant present state.There was no boundary for where she began and ended. She experienced a dissolution of her self, the boundary between herself and others was dissolved, along with a sense of deep peace and a connection with the universe. She felt like a huge ball of energy, as big as the universe, and found it wonderful to be there.

Taylor made a choice to fight to reconnect with the left hemisphere. This meant she had to relearn how to speak and understand language. It took her eight years to fully recover, and she now works on lecturing and spreading knowledge about the brain. By not letting the left hemisphere dominate the right hemisphere, she has been able to give a very clear distinction between the distinct characters of the right and left hemispheres.

Taylor describes the right hemisphere as having a very creative character. It is creative in its willingness to try anything new and be open to new possibilities. It thinks outside the framework of rules and assumptions. In the absence of rules, the right hemisphere is free to think creatively and intuitively. It is spontaneous, imaginative, and uninhibited. It processes everything into a holistic impression and experiences the world as a continuum of experiences. It thinks in wholes and takes what is present in the current moment without judging it as right or wrong. Moment by moment, it creates a collage of how each moment looks, sounds, tastes, smells, and feels. These moments are rich in thoughts, feelings, sensations, and physiological reactions.

The left hemisphere has an analytical and judgmental character. It defines boundaries and thinks within the framework of rules and assumptions. By thinking within the framework of rules and assumptions, it is limited and not open to trying anything new, but it can handle all the possibilities the right hemisphere has and shape them into something manageable. It is good at characterizing, organizing linguistic communication into something understandable. Its focus is on differences, distinguish characteristics and it is constantly busy theorizing, rationalizing, and memorizing. It thinks serially and perceives details. It thrives on details and gets stuck in details without perspective. It takes each of these rich and complex moments created by the right hemisphere and organizes them in a sequence. Through organization of details in an orderly and methodical way, the concept of time is distinguished, and moments are divided into past, present, and future.

If the right hemisphere thinks in collages of images, the left hemisphere thinks in language. It interprets and creates abstractions without regard to the whole. Through language, the world is divided into separate categories and separate objects. It uses words to describe, define, categorize, and communicate about everything. It breaks down the big picture of present moments to manageable and comparable pieces of data that it can talk about. It takes all the details accessible to work with and weaves them together into a story, thus constructing a narrative to explain everything.

The left hemisphere is designed to make the outer world comprehensible. It has the ability to quickly interpret and understand the world based on minimal attention and calculation. It is good at detecting and recognizing patterns in the outer world and has built-in recognition patterns. It interprets automatic patterns based on previously learned experiences and tends to create loops of thought patterns that are repeated over and over. The left hemisphere's pattern recognition is practically impossible to stop.

The left hemisphere is seduced by a continuous flow of thoughts. The need for consistency in this continuous flow gives the illusion of a fixed self that is unchanging and in control. By the ability to say "I am" to everything, a solid, independent entity arises, separate from the flow of experiences. It is a self that is related to the social norm we define ourselves by in relation to the outer world.

Taylor describes three important insights she gained during her recovery. The first insight was that she understood the left hemisphere's tendency to construct stories and believe them to be true. It took accessible details, filled in gaps where factual data was missing, and wove them into a story. She realized she should be vigilant of the left hemisphere's tendency to create stories and drama. By being aware of the left hemisphere's tendency to create stories and drama, she could learn not to take these stories seriously.

The second insight was that she understood she had the power to stop thoughts that caused unnecessary psychological suffering by shifting attention away from thoughts to the present moment. The third insight was that she understood that deep inside her, there was a self she experienced with feelings of joy and deep inner peace. These feelings were already programmed into the right hemisphere.

It was a self she had not previously experienced due to the dominance of the left hemisphere. Before the stroke, her personality was dominated by the left hemisphere's judgmental and analytical character. She realized it was the left hemisphere that created her personal identity, and therefore she had not previously been able to experience the intuitive and creative character of the right hemisphere.

Taylor understood that it was important to find a balance between the two hemispheres' ways of thinking. To utilize both hemispheres, we need to be aware of their distinct ways of thinking. When we understand their distinct ways of thinking, we gain the power to switch between their different perspectives. It is important to know which perspective should dominate at any given moment and occasion. Having a brain with two hemispheres that process information in uniquely different ways should increase our chances of survival.

The two brain hemispheres are separate, but they can normally complement each other. For example, the right hemisphere complements the left hemisphere's ability for verbal communication by being sensitive to non-verbal communication. The right hemisphere is tuned into subtle messages and is good at interpreting body language, emotions, and intentions. It values the entire expression in communication and assesses how the different messages being sent are aligned to each other.

It is common for the left hemisphere to dominate the right hemisphere. It is so dominant that we are not even aware of the right hemisphere's function. We can only become aware of the right hemisphere's function if the left hemisphere is not dominant. To become aware of the right hemisphere's function, we need to go beyond the left hemisphere's interpretive intellect. One area where this is particularly true is creativity. When it comes to creativity, we must rely on the right hemisphere. We need it to generate new ideas, but we need the left hemisphere to realize them.

Despite the hemispheres processing information in very different ways, they collaborate closely with each other. The two hemispheres do not share neurons, but they can exchange information through the corpus callosum's approximately three hundred million axons (nerve fibers). Apart from that, they are completely separate from each other. They are so intricately intertwined into a single unified perception that it is impossible to distinguish what happens in the left hemisphere from what happens in the right hemisphere. By being neurologically integrated via corpus callosum, they are regarded as complementing each other in a whole rather than as two separate consciousnesses and functions. They are so intricately woven together that we experience ourselves as a single person with a single consciousness.

8. DIFFERENT SELVES

Daniel Kahneman is an Israeli-American psychologist and economist reknowed for his work in behavioral economics and cognitive science, and his insight that humans do not always make rational decisions. In 2002, he was awarded the Nobel Prize in Economics for his groundbreaking work in behavioral economics. His work has significantly impacted our understanding of the irrational aspects of human decision-making, and show how beliefs influence how we take in and use information when making our decisions. He highlights our tendency to interpret information in a way that confirms our beliefs, by ignoring or rejecting contradictory information.

In his book "*Thinking, Fast and Slow*" (2012), he describes a famous experiment that uses different experiences as the basis for judgment and decision-making. The experiment consists of three parts. In the first part, which is called the short part, the participants are asked to hold one hand in a container filled with water measuring fourteen degrees Celsius. This creates an experience that is unpleasant, bordering on painful. In the second and third parts, which are called the long part, the participants are asked to hold their other hand in another container filled with water measuring fourteen degrees Celsius for one minute. Then, without their knowledge, warm water is added, raising the temperature to fifteen degrees Celsius. After thirty seconds, they are asked to withdraw their hand. Seven minutes later they are requested to choose which of the two parts they would prefer to repeat. Eighty percent of the participants say they prefer to repeat the long part because they remember it as less painful.

Kahneman concludes that the experiment shows that there are two different selves in judgment and decision-making. The interpretation and explanation for the different judgments are attributed to what he calls the experiencing self and the narrative self. The experiencing self judges and makes decisions by evaluating the experiences happening in each moment. For the experiencing self, it is impossible that a less unpleasant experience added to a more unpleasant experience can make the episode more appealing. The focus is on direct sensory experience and paying attention to the entire experience. The duration of the two parts of the experiment matters to the experiencing self.

The narrative self judges and makes decisions by averaging the peaks of moments and the final outcome. For the narrative self, it is possible that a less unpleasant experience added to an unpleasant experience can make the episode more appealing. It has focus on explaining and constructing stories, connecting various clues to narratives. The duration of the experiments two parts does not matter to the narrative self.

Antonio Damasio

Antonio Damasio is a Portuguese-American professor of neuroscience, renowned for developing a theory on how the physiological states of the body and our emotions play a crucial role in generating and evaluating options in decision-making. His work has influenced understanding of the interconnection between the brain and emotions. In his books "*Descartes' Error: Emotion, Reason, and the Human Brain*" (1994) and "*Feeling and Knowing: Making Mind Conscious*" (2021), he describes the physiology of rational thinking and how decision-making is linked to two different types of self.

One type of self is associated with our bodily sensations, that is called the core self. The other is a type of self constructed from autobiographical data, that is called the autobiographical self. The autobiographical self is a self that is constructed based on memories and derives from emotional experiences, learning, and social relationships. These memories are linked together into stories and narratives about who we are. Since we have an infinite number of memories that are connected in various ways, multiple autobiographical selves can be created. We can have many autobiographical selves, which are evoked by the environment in different ways. We can have a self created from experiences in the family that might not align with a self derived from a social circle. They can represent completely different and directly incompatible ways for an individual to act.

We can benefit from multiple personalities. Instead of trying to suppress our conflicting personalities, we can see them as a resource to use in different situations. By not suppressing thoughts that belong to different selves, we can increase our range of possible actions, enhance our psychological flexibility, and release energy. When psychologists talk about personality, they often divide it into different self-systems that operate in parallel. Sigmund Freud and Carl Gustav Jung are two of the most well-known theorists of the psyche's structure.

Sigmund Freud

The Austrian physician Sigmund Freud (1856-1939), known as the founder of psychoanalysis, developed a model of the psyche that includes the concepts of Id, Ego, and Superego. The Id represents the individual's basic drives, which can manifest in various expressions and forms. All pleasure-seeking is considered related to sexuality, which Freud called libido. Libido is the overarching drive located in the unconscious. The Superego represents an ideal self-image and adherence to societal norms. It functions as a conscience and is comparable to an ideal self. The Ego represents the consciously experienced self, influenced by drives from the Id and demands from the Superego, serving as a mediator between them. The Ego mediates between forbidden drives and what is socially and culturally permissible.

According to Freud's psychoanalysis, we experience conflicts between inner drives and moral expectations and must manage these through willpower and abilities. Drive impulses are repressed into the unconscious due to moral demands. They can resurface through the preconscious into the conscious mind in disguised forms, such as psychological symptoms. The Ego uses various defense mechanisms to prevent conflict-laden material from surfacing and becoming conscious. These are unconscious reactions that can be used to alleviate anxiety in the face of threats to the ideal self-image. These mechanisms are established already in childhood and form patterns that influence personal development and the whole societal cultures.

Freud played a significant role in Western culture during the 20th century. Many of his theories were largely abandoned during the latter decades of the 20th century, but his ideas still have followers. Psychoanalysis has gradually evolved, and remnants of the ideas developed by Freud and his colleagues still exist.

Carl Gustav Jung

The Swiss psychiatrist Carl Gustav Jung (1875-1961), known as the founder of analytical psychology, could never fully accept Freud's psychoanalysis, especially his dogmatic stance on the theory of sexuality. Jung believed there were more driving forces in humans than sexuality. With his book "*Wandlungen und Symbole der Libido*" (1912), Jung diverged from psychoanalysis and gave rise to an approach within psychology known as analytical psychology or Jungian psychology. He has a significant influence not only in psychology but also in the history of religion, literature, and philosophy. Jung developed a model of the psyche that included the concepts of the Ego, the Shadow, and the Self. These concepts are well-known and is used in everyday language, although they sometimes may have different meanings.

The ego

According to Jung, the ego is the center of the conscious psyche. It represents the central point around which all conscious mental material revolves. It is the consciously experienced self and can only perceive the conscious parts of the psyche. The ego governs our daily activities in life and gives us the sense of being separate from other people and the world. The ego includes the persona, a term derived from the Latin word "persona," meaning person, personality, role, or mask. It is the mask we wear in social settings, the roles we assume to adapt to social conventions imposed on us, and the expectations we are expected to fulfill in public life. It is the mask we present outwardly to portray an ideal image of ourselves and seek validation from others. It is a mental construct for our adaptation to social values and norms. Generally, the persona is not who we are, but what we and others believe we are.

Excessive identification with the persona indicates a personality lacking authenticity. This leads to our authentic self becoming nearly unrecognizable. Unlike the persona, which is the mask we wear when playing our social roles in public life, authenticity means being genuine and true to oneself and other.

The shadow

The shadow represents repressed thoughts, feelings, and desires that the ego does not want to acknowledge. These aspects of the ego are incompatible with our persona. The shadow contains the psychic material the ego has rejected because it does not align with social norms and the ideal image we have of our personality. Instead of occupying a place in the conscious mind, they have been banished to the shadow. This can involve forbidden desires but also abilities we possess but have not developed. It includes suppressed potential possibilities, special talents, and inherent creativity that we have not developed. When we deny the shadow, we project its characteristics onto the world and other people. The shadow cannot be denied, but it can be repressed, remaining unconscious, undifferentiated, and primitive, gaining strength. The unconscious then grows in power, forming an intrusive shadow that always follows us. The shadow can be found in all individuals, cultures, and societies.

The self

The self represents what we primarily are. It is the center of both the conscious and unconscious psyche and is the central point around which all psychic material revolves. It includes the ego and governs both our conscious and unconscious activities. The self encompasses the conscious and the unconscious, while the ego only includes the conscious parts of our personality. The self is the sum of all psychological phenomena, containing all drives, desires, motivations and needs. It includes not only the pleasant but also the unpleasant aspects.

The self is an autonomous force that can produce a flow of unconscious material, over which the ego has no control. It is a unified organizing force in the innermost being of a person and represents the goals and meaning in life. It is the source of the psychic forces that guide the our maturity and development in various ways and represents our search for becoming a unified individual. The nature of the self cannot be fully understood or identified with. Its nature can only be represented in myths, fairy tales, dreams, or symbols of wholeness.

The central process in human development and maturity is the search for becoming a unified individual, which Jung describes with the term "individuation." It is an inherent process that occurs naturally, which we undergo unconsciously in life, growing from within and flowing freely. The goal for humans is to realize the potential that lies beyond social goals. It is a process in which innate parts of the personality become integrated into a functioning whole. This can only occur through liberation from identification with the persona and our realization of the persona's fictitious nature.

Unified Self

The question of whether a unified self exists is complex. If we refer to a unified self as an unchanging core of a person, most philosophers, neuroscientists, or psychologists would likely argue that no such thing exists. In philosophy, the question of a unified personal self is one of the most fundamental questions. Philosophers have pondered how our self maintains a certain unity over time despite changes in our thoughts, body, and experiences. Some philosophers have focused on the role of consciousness in the question of a unified self, whether it is possible to have a unified self without consciousness, or whether consciousness gives the experience of a unified self. An aspect of the unified self involves the role of memory in maintaining a personal self. Remembering and constructing stories are closely related to our perception of ourselves as a unified self. Eastern philosophical traditions, such as Hinduism and Buddhism, question the existence of a permanent and unified self, emphasizing instead the absence of an enduring, independent self.

For neuroscientists, the self is a product of the collaboration between several different areas of the brain, including those involving memory and perception of the body. The sense of a unified self is a result of chemical and physical processes in the brain. The experience of a constant and unchanging self depends on construction and storage of memories. If by a unified self we mean a constant unchanging self, it is clear that no such thing exists. This is based on the idea that there is a subject within our body and brain. There is no single subject in our brain, but there are different self-systems in the brain working in parallel.

When psychologists talk about personality, they often refer to personal identity, which is a temporary structure that can change and develop through our experiences. It is tied to name, autobiographical history, and cultural background. Personal identity is constructed from memories created from words, concepts, judgments, and definitions. It makes us a person, distinguishes us from other people, and is shaped by how we perceive ourselves and how others perceive us. We associate our self with specific traits, values, interests, and roles. We identify with various groups, such as gender, ethnicity, religion, profession, or interests.

Cultural identity involves identifying with a specific culture or cultural group. It includes aspects such as language, traditions, values, and norms characteristic of that culture, influencing how we perceive ourselves and relate to others. Social identity is linked to the groups or communities an individual belongs to and identifies with. It can include factors such as gender, age, religion, nationality, and affects how we define ourselves and relate to others within the same group or to those outside the group. In our digital age, there is also the concept of digital identity, referring to how individuals present themselves digitally through the choice of images, texts, and information on digital forums and social media platforms.

Our personal identity is largely dependent on our culture and what the people we grow up with say about us. It is a bundle of thoughts, stories about who we are, what we have done and achieved, what we know, can do, and own. All these are thought forms that give us a sense of self. We identify with various things as we grow up. The name our parents gave us becomes a container for all our identifications.

We identify with people we know, our relationships, our knowledge, our abilities, and our physical strength. Over time, we accumulate opinions that become part of our sense of self, and we identify with the thoughts that constitute our opinions. We can become defensive and sometimes violent when our opinions are challenged. Throughout human history, people have been killed for not having the right opinions. In ancient Greece, Socrates was killed for having incorrect opinions. In the Middle Ages, people were killed for opinions that did not conform to conventional doctrines established by the church. In our time, people are killed for opinions that are not consistent with political regimes.

All references to personal identity are tied to our ego and the activity of our intellect. It is through identification with thinking that the ego arises. The more identification with the ego, the stronger our experience of a constant unchanging self. There are many manifestations of the ego, but fundamentally, the ego gives a sense of self. That is why we cling to it and defend it with our lives. It is culturally normal, and it has been that way for a long time. Identification with a particular tribe and seeing other tribes as enemies is example of a primitive manifestation of the ego.

Even though the ego can manifest in various ways, certain characteristics are consistent. The ego feels incomplete and is always searching for something to fill its emptiness. It constantly seeks satisfaction in future goals and expects to be fulfilled by external objects, fame, or other people. However, once obtained, this satisfaction is short-lived, and the sense of lack persists. The ego feels restless and driven by the need to achieve as much as possible, finding it difficult to be present in the moment. The present moment is reduced to a means to an end. For the ego, only past and future exist. It can only survive by continuously keeping the past active and projecting into the future. This creates a dependency on thinking, as the past provides identity and the future offers promises of satisfaction.

The ego may appear self-assured on the surface, but beneath lies vulnerability in the form of extreme sensitivity to having its value questioned. The ego constantly needs validation, control, and judge everything as right or wrong. It continually tries to defend itself, expending more energy attacking and justifying its beliefs than solving real problems. It uses explanations that are illogical, makes excuses, or denies real problems. These are strategies that infiltrate our intellect, affecting our way of thinking and decision-making. They disturb our thinking and make us blind to real problems. Our intellect is a fantastic instrument, but when we let the ego infiltrate it, our thinking becomes unclear. The focus then shifts from real problems to defending a fictitious self. Defending a fictitious self becomes more important than solving real problems.

The ego's strategies can be likened to the "fight-or-flight reaction," a general survival mechanism to instinctively either fight or flee in the face of sudden dangers that threaten physical survival. It is a reactive pattern developed over hundreds of thousands of years when we lived as hunters, and natural selection ensured that our fear reactions to physical dangers were adapted for life in nature. What was once functional in the fight for physical survival has transformed into self-defense against unreal threats directed at an unreal ego. This attitude is no longer sustainable and creates unnecessary suffering on both collective and individual levels.

The ego is not real in an absolute sense. It is no solid, independent entity, that is separated from the flow of experiences. The ego is an imaginary phenomenon, a fantasy produced in our mind through thinking. Sometimes the word ego is used to describe a person who is excessively self-centered. A metaphor for people who are very self-centered and have a high opinion of themselves is the Greek myth of Narcissus. Narcissus was an unusually beautiful youth who fell in love with his own reflection. When he came to a water source to drink, he saw the reflection of a beautiful youth in the water. He became so filled with self-admiration that he could not stop looking at the reflection. Eventually, he withered away and transformed into a flower. At the place where he died, a beautiful white narcissus flower stood. The myth of Narcissus highlights the dangers of being too self-absorbed and unable to see beyond oneself.

In psychology, the term "narcissism" refers to a personality disorder characterized by an unrealistically high opinion of oneself. Narcissists consider themselves better than others and believe they are entitled to special privileges. They depend on admiration and validation and want to be the center of attention. To protect themselves from painful feelings, they develop a grandiose self-image. There may be an apparent act of selflessness, but it always comes with the intention of gaining appreciation and praise. Relationships are used as a means to feel better about oneself, not for caring about others. According to classical psychodynamic theory, narcissism originates from weak self-esteem.

Narcissism is not the same as egocentrism, which refers to the lack of willingness or inability to consider another person's perspective. Others' ideas, feelings, and needs may be perceived as having no value, and one's own perspective is seen as the only correct one. Another word for egocentrism is self-centeredness, which refers to being excessively focused on one´s person, experiences, and feelings. It is quite common, having an egocentric perspective and behavior, and in some cases, it may be necessary. In children, egocentrism is an underdeveloped ability to imagine something beyond their own perspective.

PART II

CONSCIOUSNESS

In this part, I will address various perspectives on consciousness in psychology, philosophy, physics, theories of consciousness, phenomenology, and existentialism. I will discuss different states of consciousness related to the perception of the world, self-concept, and mental presence.

9. PSYCHOLOGY AND CONSCIOUSNESS

Consciousness is an umbrella term that covers a variety of mental and emotional states. Its exact nature and mechanisms are still subject of much research. It is a complex concept used in philosophy, psychology, and neurology, and other fields of science to describe various aspects of human experience and cognition. It can have the meaning of being aware, as having knowledge and insight of who ýou are, what you perceive, think and feel. It can have the meaning of being aware as alertness, and in mental presence.

It can encompass self-awareness, the ability to be aware of oneself as an individual, including one's own thoughts and feelings, and perception, the experience of sensory impressions from the environment, such as sight, hearing, smell, taste, and touch. It includes thoughts, the mental processes involving reflection, decision-making, and problem-solving, emotions, the subjective experience of emotional states like joy, sorrow, anger, fear, etc., and attention, the ability to focus on certain stimuli or thoughts while ignoring others.

Self-awareness and object-awareness are two distinct aspects of consciousness that describe different forms of attention and perception. Self-awareness is a specific form of consciousness where the individual is not only aware but also aware of being aware. It involves a form of reflection where the individual perceives their own existence and their own mental states. Object-awareness is the awareness of external objects and events. It is focused on external stimuli and perceptual experiences.

Self-awareness is a subset of general consciousness General consciousness should be understood as a fundamental experience of being aware, regardless of whether it concerns internal or external phenomena. It is the basic ability to experience and pay attention to something, whatever it may be. General consciousness can exist without self-reflection, while self-awareness requires a level of introspection and self-reflection.It is possible to be conscious without being self-aware. Consciousness can exist without self-awareness, as shown in the experiences of many animals and young children. Many animals and young children exhibit consciousness by reacting to the environment, but they may not have a developed form of self-awareness. They can perceive and react to objects and events without having an awareness of their own awareness.

Some philosophers propose that self-awareness can be reduced to object-awareness where the object is one's own self. This mean that self-awareness is not a separate category but a form of object-awareness. Some of the philosophers propose that self-awareness and object-awareness are fundamentally different and represent different aspects of our consciousness, while others propose that there is only one form of consciousness and that the distinction between self-awareness and object-awareness is a matter of perspective or degree rather than a difference in nature.

Object-awareness and self-awareness can be seen as two different states or aspects of general consciousness. This view recognizes that general consciousness can manifest in different ways depending on what the focus is. Object-awareness focuses on external objects and events. Self-awareness focuses on one's own inner experience and self-reflection. Viewing object-awareness and self-awareness as two different states of general consciousness acknowledges their distinct features while they can be understood within a common framework of conscious experiences.

In psychology states of consciousness is a collective term for sensory data (perception) and the beliefs, thoughts, and emotions these generate. Even things we normaly not consider to be directly created by the senses, constitute a state of consciousness, for example our experiences in dream sleep.

In healthcare, the concept of consciousness is used to describe an individual´s degree of wakefulness and the ability to use cognitive functions to interpret the world. The phrase "to be conscious" denotes a state in which sensory impressions - perceptions - reach a certain threshold value. The degree of wakefulness and attention can be lowered or heightened, thereby affecting the ability to use cognitive functions.

The term attention refers to a voluntary ability to direct our perception or mental activity towards a specific object. It is the ability to selectively focus and intentionally regulate our attention. The ability to focus develops at an early age. Children are easily distracted and focus on the most eye-catching aspects of their environment. They can focus on a specific object or task for an extended period, but they can only focus on one thing at a time and only for a short duration. It is not until school age that they can intentionally regulate their attention and take in or filter out stimuli as needed.

Our attention is not always under voluntary control. While we have our attention directed towards a particular object, we can suddenly, involuntarily notice something new. This type of attentional response arises in response to unexpected stimuli and called orienting response. It consists of sensory, somatic, and autonomic changes that facilitate sensory input to the brain. The response immediately interrupts behavior, and attention is directed towards the unknown stimulus. For a brief moment, the field of conscious attention opens up. The degree of wakefulness is enhanced, and awareness broadens to understand the meaning of the unexpected stimulus. This is a general response with an adaptive function in the face of sudden changes, and linked to evolutionary theory, that is, behavior that makes us successful in the environment we live in. It does not refer to physical orientation in space and time, but to increased alertness.

10. CONSCIOUS AND UNCONSCIOUS STATES

We are not aware of all the information processing that occurs in our brains. The majority of brain activity is related to bodily functions, including the coordination of all joint movements and the regulation of the body's internal environment. These processes ensure the various parts of the body work together. It is not the brain but the mind that is divided into conscious and unconscious states.

Conscious states include brain activities that we can consciously control. These are voluntary thought processes and motivations. There is an awareness of internal states and actions that belong to and originate from a self. It is the experience of a self that thinks and makes decisions freely. When we are conscious, we can make decisions by imagining the consequences of different choices and planning for future goals. Our behavior is not dictated by external circumstances, and we do not act reactively, conditioned by learned thought patterns. We can reflect on thoughts and feelings and do not identify with our thoughts and emotional reactions. There is a sense of mental presence and an ability to notice something new. In conscious states, we have control over actions and do not react automatically and unconsciously. We can master the impulse to act, which builds up in our nervous system, before the impulse is completed.

Unconscious states include brain activities that cannot be controlled. These are involuntary thought processes and motivations. There is no experience of a self that thinks and makes decisions freely. When we are unconscious, we cannot make decisions by imagining consequences of different choices or planning for future goals. Behavior is dictated by external circumstances, and we act reactively and conditionally, based on learned thought patterns. We cannot reflect on our thoughts and feelings but identify with thoughts and emotional reactions. There is no sense of presence, and no ability to notice something new. In unconscious states, we have no control over actions and react automatically and unconsciously. We don´t master the impulse to act, which is build up in nervous system, before it is completed. Many actions occure involuntarily without us thinking about what we are doing or being aware of what drives us.

Benjamin Libet

The American physiologist Benjamin Libet (1916-2007) conducted experiments in the 1980s that showed that before we become aware of the decisions we make, the brain has already registered our decision about half a second earlier. In these experiments, participants were instructed to note the moment they became aware of their decision by pressing a button or bending their wrist.

The results showed that the brain registered the decision with a delay of 400 milliseconds before the action was performed. Many interpreted the experiment as evidence against free will. The experience of free will was considered merely a trick performed by the mind. Libet argued that even though consciousness did not play a role in initiating a voluntary action, this did not necessarily mean that conscious will was inactive. He noted that all participants had experiences of inhibiting the execution of their actions. The conscious will could still play a role by suppressing or inhibiting actions. Consciousness could be exercised in the form of a veto, having the power to stop the actual performance of an action. The question of the existence of free will has been debated from various perspectives within philosophy and psychology.

In philosophy, the question is discussed in terms of whether free will is compatible with a deterministic universe. Philosophy uses soft and hard determinism in arguments about free will. According to soft determinism, free will is a potential ability to act differently, in the sense that an action could have been different if we had wished it so. At every moment, we have a potential possibility to act differently than we do. The will is potentially free because future events are not always determined by natural laws. Just as nature is not entirely determined by natural laws, the will is not entirely determined by natural laws. The future is not entirely predetermined, and there is a possibility to influence it.

According to hard determinism, free will is an illusion because, in any given situation, we cannot act differently than we do. We can do what we want, but at every moment, we can only want one specific thing. The will is not free because future events are always determined by natural laws. Just as nature is entirely determined by natural laws, human will is entirely determined by natural laws. The future is predetermined, and it is impossible to influence it.

In psychology, the question of free will is discussed in terms of whether human actions are driven by rational thinking or unconscious motives and drives. It uses conscious and unconscious actions in argumentation about free will. When we are conscious, we imagine that we act based on rational thinking and not on unconscious motives and drives. Motivation maintains attention on something we wish to do, and evokes action towards a desired goal. It affects the level of engagement and the perseverance during the execution of an action. If an action is an expression of intention and purpose, so is motivation an representation of intention and purpose. It provides the energy and desire to do something.

Motivation can be inspired by external or internal factors. External inspiration is outer rewards, such as financial compensation or other outward signs of appreciation, provide energy and the desire to do something. Internal Inspiration is internal rewards, such as the activity itself feeling meaningful or some other form of inner meaning, provide energy and the desire to do something.

11. ALTERED STATES OF CONSCIOUSNESS

Altered states of consciousness (ASCs) involve changes in how our senses interpret the external world. They lead to changes in world perception, self-perception, and mental presence, accompanied by alterations in our perception, cognition, and emotional experience. During an unusual experience, awareness can intensify, and the perception of the world can be heightened. These states can be temporary or lasting and can arise spontaneously or be induced by techniques or substances. Examples of altered states of consciousness include dissociation, mania, and psychosis.

Dissociation

Dissociation means separation, where the body keeps experiences and actions separate from consciousness. This condition causes disturbances in the interaction between memory, identity, perception, emotions, motor functions, and behavior. Traumatic stress is the most common cause. It is the body's way of blocking difficult memories and events. To survive, anxiety-filled memories and events must be displaced. This split is a reaction to trauma or stress that has triggered intense anxiety for survival. During a traumatic event, the person cannot integrate what has happened and is unable to react rationally.

Most dissociative disorders emerge during an unusual experience that is too difficult to handle. It is a type of defense mechanism that is temporarily disabling. There is also long-term and severe dissociation that impairs the interaction between functions such as perception, memory, sensation, motor skills, and identity, known as dissociative identity disorder (DID) or multiple personality disorder. DID involves parts of the personality separating into different personalities, with different memories, experiences, behaviors and voices, which the person shifts between. This leads to a fragmented personality, disrupting the sense of self and feeling as if someone else takes control of the body. The existence of watertight compartments between sub-personalities is shown by one personality not remembering or being able to access the others.

Mania

Mania is a mental disorder that belongs to the affective disorders. Affective disorders can be divided into bipolar and unipolar disorders. Unipolar mania refers to the individual having an altered mood alongside a normal state. In bipolar disorders, the individual can oscillate between states of mania and deep depression, with the normal state being between these mood states. To be considered a mental disorder, the mood states must be highly disabling and disproportionate to the triggering factor.

Mania is primarily characterized by an elevated mood. This euphoria is accompanied by increased energy, resulting in overactivity and a decreased need for sleep. During a manic period, the individual often experiences a dramatically reduced need for sleep and appears hyperactive. Attention and concentration are impaired, and the individual is conspicuously easily distracted. Characteristically, thought flow during these periods is marked by loose associations, meaning the person jumps from one topic to another. The individual has many ideas, thinks, and speaks quickly, and sometimes cannot be understood. Unusual connections arise between different areas, and unusual thought associations occur.

During mania, psychotic symptoms may appear, primarily in the form of delusions about one's greatness, significance, or having a special mission. Judgment is impaired, and self-esteem often becomes inflated with grandiose ideas and self-confidence.

Psychosis

Psychosis is a mental disorder involving a distorted perception of reality. A person experiencing psychosis often has difficulty distinguishing between what is real and what is fantasy, such as hearing voices that aren't there or feeling persecuted and afraid. Symptoms can appear suddenly or develop gradually over a longer period.

In psychosis, there are varying degrees of hallucinations, delusions, and thought disorders. Hallucinations involve hearing, seeing, smelling, or feeling things that are not real. The most common form of hallucination is hearing voices. These voices can sometimes be comforting but are often perceived as negative or frightening and may give orders and urge actions. Delusions involve being convinced of something that is not true. This can involve receiving messages from the radio or other people, being monitored, threatened, and persecuted. It can also involve feeling chosen, famous, or having supernatural powers. Thought disorders mean it is difficult to think coherently, to maintain a line of thought, and to complete a thought. It feels like new thoughts interrupt and disturb. Thoughts become voices that command and criticize. It can be that others can read the person's thoughts or that the person can send out their thoughts to others.

In psychosis, the intellectual ability to doubt is disabled. Only belief without the slightest doubt is a hallmark of psychoses. When psychosis is ongoing, the person is entirely absorbed by everything happening within them. What the person experiences during psychosis can be both frightening and anxiety-inducing. Sleep can become irregular, disturbed, and sometimes completely absent during the most acute phase of psychosis. There are different views on how psychoses should be treated. The common way to treat these illnesses is to alleviate symptoms with medication.

Edward Podvoll

Edward Podvoll (1936–2003) is an American psychiatrist and psychoanalyst, known for founding the Windhorse Project, an alternative initiative to support the healthy forces within a psychotic person and understand how to stimulate them to achieve recovery from psychosis. In his book "*Recovering Sanity: A Compassionate Approach to Understanding and Treating Psychosis*" (2003), he has given a summary of what is necessary to support these healthy forces in psychosis.

What makes the book unique is that he discusses psychoses without adhering to any psychotherapeutic theory, but from a deeply humanistic perspective. The book is an attempt to examine various psychotic processes under a psychological lens without preconceived theories and to depict different phases in the process. It primarily explores the individual's experiences and management of their psychosis. The book reveals some mechanisms that keep the individual in the grip of psychosis and shows what the surroundings and the psychotic person themselves can do to alleviate the symptoms of psychosis.

According to Podvoll, psychosis is a deficiency in the ability to distinguish the inner world from the outer world. This deficiency brings about experiences and behaviors that can be anxiety-laden, bizarre, or incomprehensible to those who do not know the individual well. Psychosis concerns the individual's reaction to the lack of orientation ability and attempts to compensate for this deficiency through various private explanations and projections onto reality. It is about how the individual struggles to keep their understanding of themselves and the world intact. Even during mental turbulence, there are moments of self-awareness, and consciousness. These are brief moments of alertness, offering the possibility of an observant reflective self. Moments of clarity can diminish and disappear but can also return. It is almost impossible to obliterate consciousness, but it can be obscured and made dysfunctional by circumstances.

Podvoll's book is a compilation of existing material from four individuals who have been able to describe the essence of psychosis and how they have managed to recover from it. These four individuals are John Thomas Percival (1803–1876), John Custance (1900–1990), Donald Crowhurst (1932–1969), and Henri Michaux (1899–1984). The book concludes by describing his Windhorse Project in Boulder, Colorado, which is a place where he demonstrates how his ideas can be applied in a home-like environment.

John Thomas Percival

John Thomas Percival was an English nobleman who was forced to stay in a mental hospital against his will for several years, where both his family and doctors sought to prevent his discharge. At that time, psychoses were considered incurable. Percival's illness was periodic, with healthy periods between bouts of illness. Percival suffered from depression or manic euphoria with bizarre delusions and hallucinations. He was first committed to an asylum at the age of twenty-nine. Percival recovered from psychosis during his first year in the hospital, but he was not released until two years later. Despite the conditions in hospitals at that time, he managed to achieve recovery on his own and was able to leave the mental hospital. Upon his release, he dedicated his life to sharing his hard-earned experiences that recovery was possible and how it could be achieved.

Recovery is described as a developmental process in various stages. Each stage in the psychosis carried an insight into the nature of his own psychosis. Percival had to make decisions at each step along the way, and there were many obstacles in his surroundings. The decisive moment in his recovery came when his compassion for his fellow sufferers was awakened. This gave him the energy and determination to regain his health, and he eventually dedicated his life to leading a movement that fought for the human rights of the mentally ill.

Percival saw through the mechanisms of mental illness and broke the code that bound him to psychosis. He managed to achieve recovery from psychosis on his own and kept a diary to document his return to mental health. The diary was published despite protests from medical authorities and his family.

John Custance

John Custance was an English nobleman who, after working in intelligence for four years, entered an acute manic state. Over the next twenty years, he had eight severe manic periods, followed by months of suicidal depression. During these cycles, and between attacks, he was able to describe how manic consciousness expanded and how egoistic thought patterns emerged, which became obstacles to his recovery.

Custance believed he had gained important insights during his mania. He noticed that physical sensations came long before the corresponding thoughts, and sometimes without the thoughts becoming conscious at all. He believed that in everything that happened, there were opposites, such as attraction and repulsion, positive and negative, matter and antimatter. In the psyche, there could be opposite thoughts or contradictions based on the splitting of each individual thought. In life, it was male and female, spirit and matter, birth and death.

He felt that what he learned and understood during these states was both important and irresistible, and he felt a loyalty to his experiences during manic stages of psychosis. After the manic state, he felt something was incomplete or undone and felt a need to experience the "truth" once more. He saw depression as a price to pay for this. The turning point, where mania turned into depression, was always a time of anxiety. As high as he rose, he fell just as deeply into depression. The depression was as violent as the mania; every discomfort of a feeling or thought was magnified to the extreme. He called it the universe of horrors, and in this realm, everything was the exact opposite of his manic state. While he lived in a constant sense of well-being during mania, he always felt miserably ill during depression.

Custance developed a plan to explore the wisdom of his

mania. He believed he could control the mania and worked hard to develop his mania. It was induced by his own exercises to confuse his thinking. He tried various techniques to intensify and unleash the mania. These were techniques to disrupt the rhythm of the body. He attempted to induce mania by drinking alcohol, fasting, abstaining from sleep, and being constantly active. He engaged in visual illusions and fantasies. He created images from basic visual elements such as spots, colors, and angles, then gave these images body and life until they communicated with him. Visual distortions became illusions, then visions that could be realized. Perceptual changes developed step by step, becoming interconnected and irreversible.

He describes a manic chain reaction associated with processes that were momentarily passing through his thoughts. A multitude of ideas, associations, and images from the past or future burst forth haphazardly. Thoughts were interconnected and seemed to have a life of their own, going their own way independent of his willpower. When thinking reached a critical speed, it led to a state of confusion characterized by projections that literally became animated. He experienced intense excitement and inspiration to live fully. There was a joy in letting go and feeling unlimited possibilities. There was always some pretense and impulse to act.

It was an uninterrupted intoxication of thoughts where he believed he unleashed supernormal intelligence. He was fascinated by the fact that the content of his thoughts manifested instantly. It could be anything he thought about, which then echoed in reality. In every event, he saw a favorable sign of what he should or shouldn't do next. He became hypersensitive to chance and let his life be guided by it. Chance embodied a guide that directed him.

He felt liberated from the hard shell that encased his ego and developed a form of arrogance and self-centeredness. His sense of manic openness and expansion turned into a hardening shell of the ego, trapping him in his own thoughts and making him deeply isolated and cut off from the world of humans. The freedom from shame and guilt he experienced in the manic phase turned into guilt for past sins and failures.

The all-knowing clarity during mania shifted to a sense of constant fog and darkness. All the enlightened sensations from mania were gone. In their place came gloom and disgust. The grandeur and sense of power from mania were replaced by their exact opposite. He wanted to save the world and felt responsible for all the evil and sin afflicting the world during depression. The opposing worlds of mania and depression revealed to him how utterly unreal everything in the mental life was. Each world, wise or insane, was a construction of thoughts. It was this realization that awakened him. It became a turning point where he could come out of his depression. It was the moment when recovery began.

In the 1950s, Custance wrote a book about the nature of mania and depression titled "*Wisdom, Madness and Folly*".The purpose of the book was to show that there was wisdom in madness that had not yet been fully explored. In his account of his manic episodes, he also described moments of calm simplicity and appreciation of the mundane and everyday magic. Such moments were, for him, the most authentic and cherished moments during mania. It was an appreciation of sensory impressions, glimpses of the sacred in everyday life. His last published words are found in a chapter titled "*Down to Earth*." There he wrote about the importance of a balanced state of mind and the joy he derived from everyday life. The last heard of him was that he lived at home with his family and worked hard to renovate a family farm and raise livestock.

There is a paradox in mania that when a person seeks beyond their personal identity, nothing is more unsettling than grand egocentrism. There is nothing in the elated states of madness more intrusive than their delusions of grandeur. Grandiosity is the most common of all delusions. A person on the verge of psychotic confusion tries to strengthen their personal identity by asserting themselves in an attempt to find points of orientation. What begins as a desperate attempt to find their identity ends in grandiosity. Regardless of what is real, true, or wise in the subconscious that intrudes, the delusion of grandiosity must be considered misleading.

Donald Crowhurst

Donald Crowhurst was an innovative English engineer known for engaging in various spectacular and daring projects. Typically, he had a quick intellect and an ability to overcome any doubts that he could solve any problem. He believed he could take on and overcome any challenge. It was entirely in his style, as an inexperienced sailor, to throw himself into an international round-the-world race sponsored by the English newspaper Sunday Times. He persuaded sponsors to build a technologically sophisticated boat, and on October 30, 1968, he set sail. During his solo voyage, he gradually became more and more psychotic until he finally threw himself overboard in a confused state.

His boat was later found drifting empty with logbooks, written experiences, and tape recordings describing various stages of his psychotic state. These provide snapshots of different mental states leading to psychotic self-transformation and show glimpses of temporary awakenings from his delusions. It is possible to find the boundary where the break with reality occurs. It is a boundary with a specific sequence of psychological events relevant to self-transformation. His logbooks and tape recordings describe how he experienced episodes of losing his grip on reality, feeling paranoid fear, and balancing on the edge between meaninglessness and meaning. They describe his tendency to make sudden leaps leading to delusions of grandeur and a sense of omnipotence. It is a description of a complete psychotic transformation and disconnection, documented in detail. They describe the accelerating pace of thoughts and the dissolution of the psyche.

Crowhurst was tormented by a growing realization that he would soon have to decide whether he could continue. Eventually, he decided not to make any decision and just sailed aimlessly. As he drifted without a goal, there was an increased pressure of thoughts and his attempts to manipulate the psyche. There was intense fascination and absorption in thinking, and his thinking accelerated. Quickly, a new thought came, a whole new story, and a materialization of thoughts that succeeded each other faster and faster. Consciousness became a flow of constantly recurring and disappearing mental worlds and visions. With the increasing intoxication of thinking's newfound power, new ideas and insights emerged like fireworks and revelations. A sense of total certainty emerged. It was in such a state that he claimed to have achieved freedom, but it was an intelligence running amok.

He began making fraudulent entries in his logbook. He wrote that there were problems with the steering mechanism, radio receiver, and electric generator from the start. He claimed to be in places where he was not and wrote that he had managed to break the world record for miles sailed in a day. He sent out radio telegrams but without giving an exact position. Caught in a web of deceit and full of fear of shame and disgrace for having deceived the sailing world and the press, he let his radio go silent and jumped overboard after his eight months of solitary sailing.

Henri Michaux

Henri Michaux was a French poet and artist who sought to explore and illuminate the effects of drugs and their relationship to psychotic states through experimentation. In France, there was already a long tradition of scientific researchers and artists who had experimented with various personality-altering drugs. There was also a tradition of using hallucinogens in Central America to heal the sick with psilocybin from mushrooms. Michaux aimed to investigate the nature of psychosis as revealed through hallucinogens. At the age of 57, he began to explore hallucinogens known and available in Europe. He conducted controlled experiments with mescaline, which are considered some of the most successful research efforts involving hallucinogens.

Michaux's account of his ten years of experimenting with mescaline is very detailed. He induced mescaline psychosis in himself and made systematic observations of what happened during these states. He documented his hallucinogenic experiences through essays, drawings, paintings, and poetry, describing how they related to mental disturbances. During these controlled drug-induced psychotic states, he made the startling discovery that there was no symptom or sign of psychosis that he could not witness within himself under the influence of a hallucinogen. Every conceivable mental disorder lay lurking within the depths of his psyche. Michaux concluded that his observations indicated that every person has a psychological predisposition and latent potential to become psychotic. The realization of psychosis as a constant potential evoked an instinctive fear of losing control over his thinking. Maintaining control during mental turbulence required having consciousness readily available.

Michaux prepared himself to handle the effects of the drug and braced himself to face its impact. Through yoga and meditation training, he learned to control his mind and avoid distraction. He witnessed the initial stirrings of mental illness sprouting and proliferating through his psyche as the drug took effect. It began with losing volitional control over his thinking. He could not follow through or think a thought to completion, nor could he change direction or linger on a thought. Then came his hallucinations. Imaginings became reality, and speculations turned into convictions. Everything became a sign and proof of what he already knew. Assurance turned into a conviction of the reality of what was happening. The only basis for his reasoning was a feeling of certainty akin to that in a dream state, and knowledge through direct revelation.

Next came a stage of struggle for his personal survival. Everything he had believed himself to be, his identity as the sole security and enduring reference point, was stripped from his consciousness. Every idea became something he had to contend with. The brakes in his mental system failed, unchecked thought processes could not be halted, and they revealed themselves as having a life of their own. He realized he had to function without a sense of self. Various selves came and went, each as real as the next, presenting new visions of his self. These were many new personalities, temporary and short-lived. Amidst all this, he had glimpses of the insight that there was no single self. The self was just one among a thousand other selves, always possible and always at hand. Each self felt complete but also somewhat hollow. It was transparently fabricated, and when it vanished, a temporary disillusionment followed. He understood that if one of these selves were to be held onto and elaborated, it could lead to catastrophic self-satisfaction.

Michaux's observations and experiences underline the potential for anyone to slip into psychotic states under certain conditions, particularly with use of hallucinogens. His work suggests that the mind's latent psychotic potential can be brought to the surface, revealing the thin line between sanity and madness. This awareness of psychosis as an ever-present possibility is both enlightening and terrifying, highlighting the fragility of the human psyche and the importance of mental discipline and preparation in exploring such altered states of consciousness.

12. PSYCHEDELIC DRUGS

The British psychiatrist Humphrey Osmond (1917–2004) was the first to coin the term "psychedelic" to characterize hallucinogenic drugs such as LSD and mescaline. In the 1950s, he was one of the most enthusiastic advocates for the use of LSD in psychiatric treatment. The word psychedelic can be roughly translated to something that is mind-expanding or capable of altering sensory experiences hallucinatorily. Typical psychedelic effects include changes in vision and hearing, hallucinations, mixing of senses, dissolution of the self, and confrontations with the subconscious. Through having a psychedelic experience, sensory perceptions can become confused and mixed, leading to hallucinations, such as seeing sounds or tasting colors. Sensory impressions can be greatly enhanced, leading to temporary experiences of seeing, hearing, feeling, and smelling better. For example, visual acuity can increase dramatically, and colors can appear more vivid than usual. Enhanced sensory impressions give rise to hallucinations as the brain attempts to interpret a larger amount of information than it is accustomed to. The brain uses memories and imaginations to make sense of these experiences. Anxiety and unpleasant hallucinations can also occur in stressful mental and physical conditions and environments.

One theory is that psychedelic drugs affect a type of network and areas in the brain called the "Default Mode Network" (DMN). The DMN is known to be active when a person is not focused on the external world. It is a fundamental network of interacting brain regions linked to inner self-dialogues and is responsible for trapping us in automatic loops of thinking and perception. A central idea is that psychedelic drugs can disrupt automatic network of loops, making other areas of the brain available. As the DMN becomes inactive, regions of the brain that normally work together to process information separate from each other, and new connections are established. Information that is normally processed by the DMN is then processed through other neurological pathways, causing changes in consciousness and perception.

When a signal that travels along a specific pathway in the DMN cannot do so, the brain creates new connections and crossings between neurons, finding alternative routes for exchanging information. Unusual connections arise between different areas, leading to new association in thinking. When the inner self-dialogue disappears, the brain somehow unlocks and reactivates neurons, using alternative neural pathways that haven't been used in a long time. This seems to be something the brain once knew but has forgotten and doesn't use as adults.

Psilocybin

Psilocybin belongs to the group of hallucinogens but does not produce objects or persons in the immediate surroundings, which is common with hallucinations. Psilocybin is a substance found in certain mushrooms and has been used in ancient traditional medicine in Central America. It is a drug that alters an individual's thinking and emotional state and how the brain interprets signals from the senses. Placebo studies with psilocybin indicate reduced blood flow and decreased neurological activity in the DMN. Neurons that exchange signals to process information are almost completely deactivated. The substance normally processed in the DMN needs to continue being processed elsewhere, reorganizing into new connections and neurological pathways. In the new temporary reorganization, communication between neurons in the brain is very high.

During experiments with psilocybin, a completely new perceptual level is experienced. Visual images appear with both open and closed eyes and cannot be stopped by closing the eyes. The imagination is very vivid, and complex geometric patterns and symbols are reported. In terms of emotions, experiences of love, happiness, and euphoria are reported, but there can also be experiences of unpleasant feelings and discomfort. The drug can cause paranoid experiences and severe panic attacks. A crucial factor for feeling good during an experiment with psilocybin, is being in an environment that provides comfort and safety.

The experience can be an eye-opener for the fictional nature of the personal self. A variety of shapes and identities appear and dissolve just as quickly. The drug provides insight into how the mind constantly deceives itself by creating stories that can seem even more real than the stories we have in the waking state. When the individual finishes the experiment, the individual return to a normal state with a different perspective on personal identity and what is real. This raises the question of whether our waking state is also a trick of the brain.

13. PHILOSOPHY AND CONSCIOUSNESS

The philosophy of mind is a branch of philosophy that studies the nature of consciousness, mental processes and functions, their properties, and their relation to the physical body. Within the branch of the philosophy of mind, the so-called "mind-body problem" is considered the most central question, although there are other questions about the nature of consciousness that do not concern its relation to the physical body. The French philosopher René Descartes (1596-1650) was the first to formulate the mind-body problem in its current form. Since Descartes, the question of the nature of consciousness has been a central issue in philosophy.

Central concepts within the philosophy of mind are qualia, intentionality, dualism, monism, and functionalism. Qualia are what make our experiences feel a certain way. They are the subjective qualities of experiences, such as the redness of a red object or the pain of a headache. Intentionality is the characteristic of mental states that involves always being directed toward something. For example, beliefs, desires, and thoughts are always directed toward some object, state, or proposition.

Dualism and monism are two ways to attempt to solve the mind-body problem. Dualism is the term for theories that assume there are two distinct types of reality. According to dualism, consciousness and the body are, in some categorical sense, separate from each other. There are two different forms of dualism. Substance dualism is a form that posits that mind and body are composed of different substances. The mind is an immaterial substance distinct from the material body. It assumes that consciousness is an independently existing substance. Property dualism is a form of dualism that posits that mental properties are non-physical properties arising from physical substances. Although the mind does not exist as a separate substance, its properties are not reducible to physical properties. The assumption is that consciousness involves independent properties that arise from, but cannot be reduced to, the brain.

Monism is the term for theories that assume there is only one type of reality. Only one substance exists. According to monism, consciousness and the body are not two ontologically distinct entities in contrast to dualism. There are three main forms of monism called idealism, neutral monism, and physicalism. Idealism is a philosophical view that holds that reality is fundamentally mental. The external world is seen as a construct of the mind or a manifestation of mental activities. It assumes that the only thing that exists is consciousness, and that the external world is either mental itself or an illusion created by consciousness.

Neutral monism is a theory that posits a neutral substance that is neither mental nor physical, with both mind and matter being derived from this fundamental substance. It assumes that there is some other, neutral substance that is neither mental nor material, and that both consciousness and matter are made of this unknown substance. Physicalism is a view that posits that everything that exists is physical or dependent on physical processes. In this view, consciousness arises from and is entirely dependent on physical states and processes in the brain. It assumes that the only existing substance is material, or that consciousness is part of the material, physical world. According to functionalism, a mental state can be identified by what it does rather than by what it is made of. It assumes that they are defined by their functional roles rather than by their physical composition.

Many philosophers of mind have adopted reductionist or non-reductionist physicalism, maintaining in different ways that consciousness is not something separate from the body. Reductionist physicalism assumes that all mental states and properties will eventually be explained by scientific theories about physiological processes and states. Non-reductionist physicalism assumes that, although consciousness consists of nothing more than the brain, the words we use to describe mental states cannot be reduced to physical explanations. A mental state is not the same as a physical state, nor is any mental state solely composed of physical states and phenomena. Why our mental states have subjective qualities (qualia) and can be directed toward something (intentionality) remains unanswered in naturalistic terms.

The mind-body problem continues to be a significant philosophical challenge. How can mental experiences arise from physical brain processes? Can consciousness be fully explained by neuroscience? These questions highlight ongoing debates and the complexity of understanding consciousness. Despite advancements in cognitive science and neuroscience, the subjective nature of experiences and the unique characteristics of consciousness is puzzles for philosophers and scientists.

14. PHYSICS AND CONSCIOUSNESS

Physics is a science that studies how nature works at the most fundamental level. Originally, the word "physics" was a collective term for all natural sciences, but over time, it developed into the distinct disciplines that exist today. Physics deals with the components that make up matter and the fundamental forces that influence these components. It conducts studies on all scales, from the smallest subatomic level to the largest scale of the universe within cosmology.

In physics, consciousness is considered something metaphysical and therefore not a subject of physics. Physics explores the microcosm and the macrocosm but chooses to disregard consciousness because it cannot be observed through a microscope or a "macroscope." Physics assumes that the brain is necessary for consciousness to arise and considers consciousness to be a result of activity in the nervous system. Physics has attempted to discover the smallest particles but has no explanation for how consciousness emerges from matter, nor any explanation for the substance in which material objects arise. Most researchers agree that there is a connection between the brain and consciousness, but opinions differ on how consciousness is related to brain activity.

There are different perspectives on the relationship between consciousness and the brain. One perspective is the assumption that consciousness causally interacts with the brain. Brain activity is the immediate cause of consciousness. Changes in mental activity cannot occur without being accompanied by changes in brain activity.

Another perspective is the assumption of a hierarchical dependency between consciousness and the brain. This means that something additional and unpredictable arises that cannot be foreseen or calculated from properties at a lower level. When matter is organized in an appropriate way, so-called "supervenient" properties arise. This means that "layers" with increasing complexity are formed, where each layer has its own special way of being and its own science, and must be studied in different ways. For example, water has a new property when hydrogen and oxygen combine. This is a new property that could not have been predicted by fully understanding the nature of hydrogen and oxygen.

A third perspective is the assumption that consciousness is a field that transcends the brain. Even though the brain and consciousness are connected, this does not exclude the possibility of consciousness being related to a "field" that extends beyond the brain. Consciousness can be extended beyond the brain in the same way that a magnetic field extends beyond the magnet. Just as a magnetic field cannot be decomposed, consciousness is a field that cannot be decomposed. In physics, a field is an invisible force that does not have properties or act in the same way as material objects.

Evolutionary Psychology and Neuropsychology

Evolutionary psychology and neuropsychology are two important research areas in natural science. Evolutionary psychology focuses on exploring the evolutionary causes of innate behaviors and mental abilities. It is applicable to all organisms with a nervous system, but most research in the field focuses on humans. They are living organisms and should therefore be understood as part of biology. Just as evolutionary physiology aims at identify physical adaptations of the body, evolutionary psychology aims at identify emotional and cognitive adaptations.

Evolutionary psychology seeks to investigate how behavioral patterns, personality traits, and cognitive abilities are genetically conditioned. It deals with how genes have become more or less common depending on whether they provide reproductive and survival advantages through natural selection. Human nature is assumed to be a universal collection of evolved psychological mechanisms specialized in solving adaptive problems that have recurred over long periods.

Research in evolutionary psychology is focused on identifying the evolutionary advantages of different mental abilities and the selection processes that cause these abilities to spread among humans. It analyzes the causes of abilities related to vision, hearing, memory, motor skills, and cognitive functions. Evolutionary psychology is based on the idea that our brain has evolved to generate behaviors that adapt us to our environment. Different neural circuits in the brain are specialized in solving various adaptive problems. The neural pathways have been shaped through natural selection to solve the problems our ancestors encountered as they evolved into "homo sapiens." We live with a "stone age brain" that is still not fully developed.

Neuropsychology explores which parts of the brain that change their activity during different forms of mental activity. The goal is to gain knowledge about the brain areas that underlie our mental abilities. Neuropsychology is focused on how different parts of the brain affect behavior and which parts control various functions. The studies can be divided into two main areas: studies focused on the functional organization of the normal brain and studies focused on individuals with various diseases or injuries affecting brain function. The relationship between human behavior and brain function is studied from different perspectives and under different conditions. This can include states of sleep or wakefulness when a person solves various types of tasks, which can be practical or emotional in nature.

The methods used in neuropsychology have undergone rapid development since the 1930s. It began with the recording of the brain's electrical activity using electroencephalography (EEG). New techniques have since been developed, such as computed tomography and magnetic resonance imaging, which have made it possible to determine brain activity with greater precision. With techniques that measure metabolism and blood flow in the brain, it has become possible to analyze the brain in detail to find out which parts of the brain change their activity during different types of mental activity. It is possible to get a snapshot of the brain's work during various sensory impressions, movements, different types of emotions, and ways of thinking. Studies have shown that humans use different facial expressions for emotions such as satisfaction, surprise, and anger. Higher mental abilities are more complicated and have not been traced to any single specific area in the brain. Instead, they result from activity in several different centers spread across large parts of the brain.

Research has been able to demonstrate correlations between mental events and neurological events, but observing different brain states is not enough to explain consciousness. However close these correlations may be, they do not justify disregarding the experiential side of reality. Neurological events are processes of information and computation. Mental events correspond to neurological events, but they have distinct subjective qualities. The existence of cerebral events alone cannot explain why they correspond to qualitative experiences. Many cerebral processes can occur simultaneously with qualitative experiences, but this does not explain why these experiences arise. The question of why many cerebral processes are accompanied by corresponding qualitative experiences remains unanswered.

Although it is possible to demonstrate neurological correlations between consciousness and the brain, there is no theory to explain this. Neither evolutionary psychology nor neuropsychology has an explanation for why conscious experiences arise from neurological activity. There is no plausible explanation for how consciousness can arise from a lump of gray matter that consists of electrochemical properties. No one has seen how neurological processes in the brain can produce consciousness. Mental events and neurological events are different, but both are needed to describe reality.

Many philosophers distinguish between consciousness that stems from access to information in the brain and phenomenological consciousness. Phenomenological statements say nothing about neurological events. We can make phenomenological statements instead of describing mental events as neurological events. The American philosopher Thomas Nagel says that a system is phenomenologically conscious if there is something it is like to be that system; an organism has a conscious state if there is something it is like to be that organism. Phenomenologically, existence is experienced as having an "inner" mental aspect, distinct from but interacting with an "outer," directly observable, spatial physical aspect. We can feel that there is a difference between a conscious and an unconscious state, but we cannot know what mechanism gives rise to consciousness. We can only know what a conscious state is through introspection.

The Australian philosopher David Chalmers argues that subjective experiences have a physical basis, but we must explain why there is something it feels like to have subjective experiences. Mental events cannot be reduced to information processing and computational processes. If we try to explain mental events in a reductionist way, we must explain why mental states are experienced in a particular way.

The easy problem is to explain the mechanism behind subjective experiences. This involves describing and understanding neurological processes that correspond to our cognitive and behavioral functions. It is about finding a neurological or computational mechanism that performs the perceptual processes and the brain processes that lead to these behaviors. It is possible to gather data on how we perceive, think, make decisions, and behave.

These are objective things that science can measure, and they can explain the function of consciousness as purely mechanical processes in the physical world.

15. THEORIES OF CONSCIOUSNESS

There are several different theories of consciousness. Most theories are developed mathematically or intuitively and are difficult to prove. Many are based on neuroscience and attempt to explain consciousness as neurological events in the brain. There is yet no scientifically accepted theory of consciousness, but there are various attempts to explain the experience of consciousness. Five researchers who offer different explanations for consciousness are Max Tegmark, Giulio Tononi, Roger Penrose, David Bohm, and Donald Hoffman.

Max Tegmark

The Swedish-American physicist and cosmologist Max Tegmark argues that consciousness is like a pattern of information. It is a particle arrangement that has properties beyond its particles. It is the same particles, but transformed into a new pattern. Information processing gives rise to phenomena at a higher level of consciousness. It is like a process where a complex pattern is formed from the interplay between simpler structures or behaviors. When we are in different states of consciousness such as REM sleep, deep sleep, and wakefulness, it is the same particles, but it is not the particles themselves that make the difference in these states. When we fall into REM or deep sleep, consciousness fades due to the rearrangement of particles. It is the structure of information processing that matters, not the structure of the matter that performs the information processing. If physical reality is entirely information-based, no aspect of consciousness is beyond the reach of science.

Consciousness can be made of something as simple as particles because it is an emergent phenomenon with properties beyond its particles. It is like solid matter, liquids, and gases, which are all emergent phenomena. They are more than the sum of their parts, as they have properties beyond those of their particles. An example of an emergent phenomenon is wetness. A drop of water is wet, but an ice crystal and a vapor cloud are not, even though they consist of the same water molecules. This is because the property of wetness depends only on the arrangement of the molecules. A water drop, ice crystal, and gas are all made of the same water molecules, but it is not the molecules that make the difference. There is no meaning to say that one single water molecule is wet, because the phenomenon of wetness only arises when many molecules are arranged in the pattern we call liquid.

Tegmark describes consciousness as a physical phenomenon that feels non-physical because it is like waves and calculations. If consciousness is the way information feels when processed in certain complex ways, consciousness must be substrate-independent. Just like waves, consciousness can use any substrate. Waves have properties like wavelength, frequency, and speed and can be described in mathematical equations. Similarly, consciousness is independent of the substrate and can be described in mathematical equations.

If it is possible to make predictions about which brain processes are conscious and which are not based on a theory, these predictions could also be applied to other arrangements of matter. The hard problem is to show which physical systems are conscious and which are not conscious. Although it is possible to make predictions about phenomenology and its neurological correlates, it is not possible to make predictions about consciousness in machines (Tegmark, 2018).

Giulio Tononi

The Italian psychiatrist and neuroscientist Giulio Tononi argues that it is not enough to explain consciousness as a pattern of information. Information processing must also be integrated. Tononi offers a mathematical theory of consciousness known as "Integrated Information Theory" (IIT). According to Tononi, consciousness is measurable with a mathematical measure he calls Phi. Phi is a measure of integrated information, in other words, how well different parts of a system are integrated. By using Phi, it has been possible to identify whether a person is awake or dreaming, however not during deep sleep or anesthesia.

Tononi assumes that consciousness is the basis for physical objects and events. He highlights certain requirements that must be met by physical systems to serve as a basis for consciousness: conscious systems must be able to store and process information, they must be independent of the external environment, they must not consist of independent parts, but form an integrated whole (Tononi, 2012).

Roger Penrose

The British mathematician and theoretical physicist Roger Penrose argues that consciousness cannot be explained or reduced to computational processes. Consciousness lies beyond the laws of physics, computation, and artificial intelligence. Consciousness does not originate from complex computations performed by cerebral neurons. Transformation to consciousness does not occur at the level of the brain's neural circuits. Consciousness originates from a quantum level inside neurons rather than being a product of connections between neurons. Consciousness is a quantum process organized by tiny tubular cellular structures called microtubules. Cellular microtubules are the only place where a non-computational process occurs in isolation. Through the collapse of the wave function in cellular microtubules, consciousness emerges undifferentiated and unbound. The collapse of the wave function is random, and no computational process or algorithm can predict the outcome. Consciousness first enters the microtubules and then becomes intertwined with the brain's neural circuits.

David Bohm

The American-British quantum physicist David Bohm (1917-1992) argues that consciousness is a whole with correspondence in outer physical manifestation. We cannot understand consciousness by studying only the outer physical reality. He introduces the concepts of explicit order and implicit order. These are terms for different orders of the same reality. In the explicit order, space and time are the dominant factors determining the connection between dependent or independent elements. The explicit order involves abstractions that we normally perceive. The implicit order is a deeper and a more fundamental order of reality than the explicit order. It is an unmanifested order that precedes the manifestation of an explicit order.

Bohm considers each event as a prognosis or projection from the implicit order. Our usual concepts of space and time, along with the material particles, are abstracted forms derived from the deeper implicit order. What we take as physical reality are forms that evolve from the underlying implicit order. It is from an implicit order that the universe's explicit order arises. Movement, flow, and change of experience are all manifestations of it.

Donald Hoffman

The American cognitive psychologist and AI researcher Donald Hoffman proceeds from the assumption that consciousness is foundation for objective reality. All theories must be based on some assumption that is taken for granted. Since a theory should be based on as few assumptions as possible, it is better to start from consciousness as a self-evident assumption than to start from material theories based on more assumptions than a theory based on a single assumption.

Instead of starting from matter, Hoffman starts from consciousness as the fundamental basis of reality. He uses mathematical models and studies consciousness from the perspective of visual perception, conducting several psychophysical experiments. He creates a mathematical model that encompasses all the different aspects of consciousness. It is a model of consciousness with testable hypotheses. Hoffman states that the model is only valid if it can derive relativity theory and quantum theory since these theories have significant evidential value. The model is based on what he calls "conscious agents." He makes a mathematical definition of these and can calculate possible perceptions and actions for different conscious agents.

He uses geometry to describe how conscious agents compress information into something that can be seen and shows that when two conscious agents come together, they form a new conscious agent. Conscious agents can interact with each other and form a complex network of conscious agents interacting with each other. They can combine to create more complex agents. It is a process that leads to an infinite number of agents with infinite potential for experiences, decisions, and actions.

It is a theory where consciousness involves perceptions, decision-making, and actions that influence the world in feedback loops. Everything can be transformed into mathematical symbols, and it is mathematically possible to specify attention, memory, problem-solving, and learning. The inner world of mental experience has a mathematical structure, and this mathematical structure is the basis for describing both the physical world and the inner mental world.

In his book "*The Case Against Reality*" (2019), he describes his theory and his research on consciousness. Hoffman uses evolutionary game theory, which is a mathematical theory applied to natural selection. He asks what the likelihood is that natural selection could create sensory systems to see the truth about objective reality. The answer is that the likelihood is zero. Every sensory system in organisms has not been created to see the truth about objective reality.

Evolutionary theory is about competing strategies for survival and reproductive fitness. To know something about the truth is not helpful for reproductive fitness. The purpose of evolution is not to shape our perceptions to see the truth about reality. Through logical reasoning, experiments, and mathematics, Hoffman shows that our senses do not reveal true reality. What our senses show us is an interface to objective reality, analogous to a user interface on a computer desktop. Evolution has given us our senses as an interface to interact with objective reality.

Our perceptions are designed to hide reality. They are focused on usefulness for survival, not for truth. They operate according to evolutionary laws of adaptation and have nothing to do with reality. We are prone to falsely believe that our perceptions provide genuine insights into an objective reality, but we perceive reality through an illusion created by our perceptions. There is a reality outside our brain and body, but we construct reality based on usefulness. Every physical object we see is a representation we have in a virtual reality, and we take that virtual reality to be objective reality.

Objects, shapes, space, and time reside in consciousness. If we did not exist, none of these qualities would exist either. Hoffman uses icons on a computer desktop as a metaphor for how our perceptions work.

Icons on the computer's desktop represent only usefulness, not any objective reality. The computer's icons are symbols used to hide the truth about diodes, resistors, voltages, and magnetic fields. We do not need to know about all the circuits that exist. We can control them without knowing the truth about them. These icons are not there to imitate the truth; rather, they are there to hide the truth that exists inside a computer.

The icon for email can be blue and rectangular, but that does not mean the email is blue and rectangular. The purpose of an interface is not to show us the truth about the computer, where truth in this metaphor refers to circuits, voltages, and layers of software. Rather, the goal of an interface is to hide the truth and show a simple graphic that helps us perform useful tasks, such as creating email and editing photos.

Science has long been based on the assumption that spacetime is fundamental to objective reality and says nothing about what lies beyond spacetime. It has been exceptionally successful, but the theory of natural selection and quantum field theory say that spacetime is not fundamental. Most physicists admit that spacetime is not fundamental, even though it is difficult for natural scientists to accept because they assume that consciousness originates in the brain, which is perceived to exist in spacetime.

Hoffman argues that spacetime is not fundamental to objective reality. What we perceive as spacetime is an "interface," a computer desktop that hides spacetime. Spacetime and its physical objects exist only as created by us. It is erroneous language to use spacetime for experiences of shapes, colors, and physical objects. Physics and evolution point in the same direction: space, time, and objects are not fundamental. Spacetime emerges from something else that is fundamental. Spacetime is just a data format that serves to keep us alive. Spacetime is our virtual reality, a "headset" of our doing.

Metaphors

If consciousness cannot be explained with a theory, only metaphors remain. One metaphor for consciousness is the ocean. Each wave on the ocean is unique and different from the others, but the wave is still just a movement in the ocean. Consciousness is all-encompassing at the ocean's level, and the wave on the ocean is the individually conscious. At the ocean's level, everything is interconnected, and thus all differences disappear. What is described as the "conscious mind" in analytical psychology can be identified with the top part of the wave. Everything else is subconscious, and deep down in our subconscious, there is a level where we are all connected in what is called the collective unconscious. When we live our lives with attention only to the external, we experience only the top of the wave, and everything else goes unnoticed.

Another metaphor is the sky. Consciousness, like the sky, is not a thing or object. In the same way that the sky is the unmanifested background for everything that exists appears, consciousness is the unmanifested background for everything that exists and appears. The unmanifested is beyond the outer visible world, with all its physical objects in spacetime. It is timeless and formless, and we cannot imagine or intellectually understand it. What is manifest is the outer visible world, with its physical objects in spacetime. It is time-bound and form-bound, and can be imagined and intellectually understood. This includes the world of time and separate forms.

We notice the stars but not the space in the universe itself. We would not notice the stars if there were no space. Without space, there would be no stars. It takes more than one reference point for distance and space to exist. The universe would not exist without space, and yet space is nothing. Before the universe came into being, before the "big bang," there was no empty, vast space waiting to be filled with things. There was no space because there were no things. If there were only space without things, the universe would not exist for us. There would be no speed, or movement from one place to another.

Space has never been created. It is only when there are multiple entities that the manifest world arises. Everything that exists has its origin in a vacuum state. If there were no vacuum, no objects would exist. The vacuum state is unimaginable, yet it underlies everything in this world. All material, physical forms originate from a vacuum state. Even what appears to be solid matter originates from a vacuum state. The vacuum state is more like a vibration frequency than particles of solid matter. It vibrates with life, and life in every form. The density of matter is an illusion. According to quantum field theory, it contains fleeting electromagnetic waves and particles that pop in and out of existence.

16. CONSCIOUSNESS AND THE PHYSICAL WORLD

Anil K. Seth is a well-known British professor of cognitive and computational neuroscience. Seth questions how consciousness is related to the external physical world and how we can understand the relationship between the inner world, which consists of our subjective experiences and sense of self, and the physical body. We perceive the world and ourselves through and because of our bodies. Consciousness is the subjective experiences that an organism can have, and it has a close relationship with the physical body. Consciousness is often associated with being awake, but it is not the same as being awake. It is possible to study how conscious we are in different states of wakefulness and study different levels of consciousness. A difference in brain energy consumption can be seen when we are awake and asleep, although it is small.

Consciousness is, like life, something beyond what can be explained as mechanisms. It is not possible to say that consciousness is only about mechanical processes in the brain because consciousness is the space in which all experiences appear. Robots have no consciousness, no reflective self, or self-awareness. A computer that can play chess or go is not conscious. Consciousness cannot be explained as mechanical processes, but we can ask what we do when we are conscious, says Seth.

Seth describes consciousness in terms of information and integration. It is not the absence of information or integration itself that measures a level of consciousness, but how information and integration are jointly expressed. Both information and integration are required for the emergence of a conscious state. A brain with a certain amount of information but no integration, or a brain with integration but no information, would not support any conscious state. In a maximally information-rich brain, without integration and order, all neurons would be independent of each other, and the behavior would be completely incoherent. In a maximally integrated and ordered brain, all neurons would be dependent on each other and do exactly the same thing.

Conscious experiences are informative. Integration is an information-producing property in our conscious experiences. Every conscious experience is different from any other experience we have had, will have, or could ever have. Every moment has a conscious experience from an infinite number of possible experiences.

Every specific experience is defined by unrealized possibilities it does not have. For example, experience of the color red is not due to inherent property of redness. It is red because red cannot assume any other color. The same applies to the other conscious experiences. It can be a thought, a feeling, or any other form of mental content. Conscious experiences are not only highly informative, but also integrated and are not separate from the external world. This means that every conscious experience appears as a unified scene. We do not experience an object independent of its background. The different elements of our conscious experience are what we currently experience with our senses, thoughts, and feelings. The goal of perception is to promote our survival.

We do not perceive the world as it is but as it is useful to us. Our perceptions are projections of our brains. They are structured expectations seen from an observer's eyes. They only seem to be objective, realistic properties of an external world. The property of appearing real adds extra fuel to our dualistic intuitions about how conscious experiences and the physical world relate. The self is another perception, a controlled hallucination, but of a special kind. It is the experience of embodied personality that relates directly to the body. Emotions and mental states are also aspects of embodied personality. The sense of self is a collection of perceptions, a tightly woven bundle of neurologically coded predictions aimed at keeping the body alive.

Our perceptions change all the time, but we do not notice it. We do not see things as they are; we see them as we are. Our conscious experiences of the world around us and of ourselves within it happen with, through, and because of our living bodies. The deepest level of the experience of personality, the sense of just being, lacks external references altogether. The same happens with our experience of being a self. Our mind and brain are evolutionarily designed to perceive ourselves as less changeable than we perceive ourselves to survive. This subjective blindness promotes the false intuition that the self is an unchanging entity rather than a bundle of perceptions. Seth calls this "self-change blindness".

Pontus Wasling

Pontus Wasling is a Swedish brain researcher and physician known for his research on the function of sleep, neurological diseases, and the smallest components of memory. In his profession, he meets patients with diseases that in various ways affect the brain. In his book "Knaster: medvetandets gåta och illusionen av ett jag" (2021), he tells how one of his patients woke up from a coma but was still not conscious. After a few days of coma, the person woke up in the hospital and quickly fell into his usual routines. He looked and acted normally but was not really present, nor completely absent. The body's mechanisms worked as before, but there was no consciousness or sign of the experience of what we call self or soul. Wasling asked whether the body could function without consciousness and whether it was possible to be conscious and have a sense of self without a body.

It has not yet proven possible to be conscious without a body, but there are cases that show that it is possible for a body to function without consciousness, which can be caused by a severe head injury or severe senile dementia. These are people who only eat, sleep, and function basically on a bodily level, a so-called vegetative state. They are alive, but what we call consciousness is absent. There is no sign of self-experience, the ability to feel pain, or any goal-directed activity. They lack human intellect and emotional life.

There are people who can be conscious but lack the ability to communicate with the outside world. Jean-Dominic Bauby (1952-1997) was a French journalist, author, and editor who, in 1995, at the age of 43, suffered a stroke that resulted in the so-called "locked-in syndrome." When he woke up from a coma after twenty days, his entire body was paralyzed. Bauby could only move his left eyelid and developed a way to communicate by blinking. Despite this, he was able to write his memoirs through blinking. In his book "*The Diving Bell and the Butterfly*" (1997), he writes about the isolation and frustration he felt. Bauby died ten days after the book's publication. The book was later made into a film in 2007 under the same name

17. THE ORIGIN OF LIFE

Paul Davies

Paul Davies is an English physicist, cosmologist, astrobiologist, and author of hundreds of research articles in various scientific fields. In his books "*God and The New Physics*" (1983) and "*The Mind of God*" (1992), he poses the question of what life is and how life can arise from non-life.

Metabolism and life are prerequisites for our existence and consciousness. It involves taking in and transforming energy into something useful for the body's physical survival, meaning interaction with the environment to sustain life processes. The production of new living organisms is called biogenesis. The law of biogenesis is the observation that all living things arise only through the reproduction of other living creatures. This view has dominated for the past hundred years.

In biology and chemistry, there is the idea that life can develop by bringing together the right chemicals. In laboratories, experiments are conducted to try to take the first steps towards creating building blocks with amino acids in a simple way. The problem is obtaining the right kinds of molecules that are sufficiently complex. This idea rests on the assumption that the early Earth had complex chemicals that organized themselves into a living entity.

Davies argues that it cannot be as simple as a random organization of complex chemicals creating life. There is no evidence to suggest that it is an easy transition from non-life to life. It is impossible to estimate how likely it is that a mixture of chemicals will transition into something as complex as a cell. The view that life arises through a mixture of chemicals relies on the belief that there is a deep life principle in the laws of nature that gives matter a fast track to life. Both biologists and chemists have tried to create life in laboratories, but they do not know how chemicals transition to life. They do not know the physical process for the transition from non-life to life. If we do not know the mechanism, we cannot estimate the probability of life arising.

Biologists talks about life in DNA, with amino acids making peptides and proteins. The primary task of peptides is to transfer information to the cells on how to function optimally. Two things happen in DNA. One is that the base pairs in genes can read instructions and get the protein to operate in a certain way. The other is that the gene pair can become a packaged and replicated object. It replicates not only as an object but also as instructions that follow the object. What to do with DNA, and when to do it, is programmed into the cell.

Chemists speaks of substrates, which are the substances in which life exists. This means that the problem of the origin of life concerns hardware, but the real problem of the origin of life concerns software. It involves organizing, managing, and storing information. What makes life seem different from any non-living complexity is the way it stores, organizes, and manages information. The first step for organizing and managing information has to do with the replication of instructions. There are laws that govern physical processes and there must also be laws that govern the flow and organization of information. This does not necessarily mean that it infringes on the laws of physics, but only that they cannot be reduced to each other, says Davies.

Natural science can only explain the universe as a physical mechanism, but it says nothing about why we exist and are conscious beings. The question of why our universe exists and why humans are conscious witnesses to the universe are existential questions that natural science has no answer to. It can explain how all parts of a clockwork mechanism function but not why the clock exists. Consciousness is our most indisputable fact about what is real, yet it has been entirely overlooked in natural science. The idea that consciousness could play a fundamental role has been excluded from natural science because it is impossible to detect any particles of consciousness. Physics has many taboos against studying consciousness, but consciousness may have a deeper reality than previously imagined. To understand consciousness, a deeper understanding of it is required.

Natural science seeks initial conditions and causes for the existence of everything. Every time a new initial condition is proposed, something else that requires explanation is added. This leads to an infinite regress where only something else that needs explanation is continuously added. To avoid the problem of infinite regress, a fundamental base of reality that is self-founding is required. Consciousness is not grounded in anything other than itself and it is the only level of existence that can provide an explanation that is self-founding. Consciousness can be the absolute foundation of existence. Existence has no rational explanation. It relies on the assumption that we understand something only when it is logically and rationally explained. Rational explanations are not enough to understand the existence and properties of the universe. Equating understanding with rational explanations leads to an infinite regression.

There is a limit to knowledge through rational thinking. A rational explanation of the world, in the form of a closed, complete system of logical truths, is not possible. The very rules of logical reasoning force us from the outset to seek an explanation of causes. If we want to go beyond this, we must embrace a different concept of understanding than what rational explanation is, and there is another way to knowledge beyond the path of rational questioning and logical reasoning. Many of the famous researchers did not make discoveries solely through rational thinking. They had mystical insights that led to the formulation of new scientific theories.

According to Davies, it is meaningless to ask for any cause of the very first event when time and space came into being because every cause must be sought within time and space. Physicalism can never provide a sufficient explanation for the existence of our universe because there are problems with infinite regress. Some argue that the underlying nature of the universe is more akin to consciousness. Quantum physicists describe the ultimate foundation of reality as particles that come in and out of existence. A deeper explanation than what quantum physics describes is required. There is no reason why the ultimate foundation of reality should be limited to the level described by quantum physics. Quantum physics and physiological laws may have their roots in consciousness.

In his book "*The Goldilocks Enigma: Why Is the Universe Just Right for Life*" (2007), Davies writes that the laws of nature are remarkably and wonderfully well-suited for the development of life, living beings, and consciousness. They are too good to be the result of a simple random selection. Just a slight change in the laws of physics would result in the elimination of all possibilities for conscious life. Davies concludes that the successful, harmonious, and resourceful laws of physics in the cosmos are a sign that there is purpose and meaning in the universe.

The reductionist approach cannot explain everything. An atom can only affect atoms in its immediate vicinity. It can only act locally. The question then is how a liver cell knows it is to become a liver cell or how a fertilized egg knows it is to become a human. These are fundamental questions that must be answered but have been ignored. Research has paid too little attention to the problem by simply defining it away. Consciousness can never be explained by the interaction of individual atoms. Each level of complexity has its own laws. This does not mean that we must reject the laws for how atoms interact but that new ones are needed to describe a higher degree of complexity.

18. IDEALISM AND PHYSICALISM

Idealism and physicalism are two philosophical orientations with conflicting views on what constitutes the absolute foundation of existence. According to idealism, consciousness is the only existing substance. It is the primary foundation of existence. The physical world is created by consciousness and dependent on underlying mental structures, forms, and ideas. It is a world that only contains mental things and exists depending on our perception of it. The world is a product of consciousness, perceptions, representations, and conceptions. Within idealism, there are two variants: subjective idealism and objective idealism.

Subjective idealism holds a personal, subjective consciousness as the foundation of existence and comprises our perceptions, representations, and conceptions. Subjective idealism does not recognize that there is a real objective world beyond our minds. It denies the existence of real, independent entities that affect our senses and perceptions. The physical world is merely our representations and the mental processes that appear on the screen of our perception. The world we perceive through the quality of our perceptions is irreducible. Consciousness, the mind, is at the bottom of all reduction.

Objective idealism holds a universal consciousness or a universal self as the foundation of existence, which exists independently of humans. Objective idealism does not deny the existence of real entities independent of humans that affect our senses and evoke perceptions in us. It recognizes the existence of a real objective world beyond our minds. The objective world, like the human mind, is mental but not accessible to us. The entire world that is fundamentally mental processes.

The idealistic thought can be found in both Eastern and Western cultures. The roots of idealism can be traced back to ancient Vedic scriptures thousands of years ago. Idealism emerged in the West during antiquity and was particularly developed by the Greek philosopher Plato (428-348 BC). At the center of Plato's philosophy is the theory that ideas exist in an abstract world beyond the sensory world. What our senses perceive is not the real world. It is merely a shadow image of the real world. Plato claimed that the real world was the world of ideas, while the world we see is a world of shadows, a world of pale reflections from the ideas. The real world exists beyond space and time. There, everything is perfect and eternal. All forms, templates, or origins exist there and they are not accessible to our senses.

The reality we experience with our senses is temporary and imperfect. For example, a drawn circle exists in our perceived reality, but it can never be perfect. All drawn circles share the commonality of referring to the perfect circle. The perfect circle is an abstraction and does not exist in our world. It exists outside time and space and exists only as the idea of the circle. The idea of the circle is non-material and eternal and is in its definition perfect. Everything that exists in our world is inferior copies of a perfect abstraction. If we look a little closer, we can always find some defect. Humans live in a body with incomplete organs of knowledge. It changes, grows, ages, and dies, but humans as an idea remain the same.

Physicalism (Materialism)

According to physicalism, the only existing substance is matter. Matter is the primary foundation of existence. The physical world is a creation of matter, and it contains only material things that exist independently of our perception of it. Consciousness and mental content are products of matter and the brain. Everything originates from physical causality in the brain and is described in mathematical formulas and algorithms. Everything we remember, all the skills we have learned, are stored as material traces in the brain and follow physical laws. Consciousness is merely a byproduct of brain development or an illusion created by the brain.

Physicalism has its roots in classical atom theory. Classical atom theory is a line of thought that comes from the Greek philosopher Leucippus and his student Democritus. According to classical atom theory, reality is made up of an infinite number of indivisible units, so-called atoms, which have always existed in different forms and sizes. These atoms are invisible, colorless, and solid material particles that follow the laws of mechanics and interact with each other through collision. Classical atom theory developed during antiquity and became the starting point for modern atom theory and the development of physics, which took off from the 17th century onwards. Physicalism took a dominant position until the breakthrough of quantum mechanics in the 1920s.

Physicalism has a mechanical and deterministic view of the universe, nature, and humans. Everything in the universe happens mechanically, unconsciously, and without intelligence. Everything in nature is made of matter and is governed by unchanging laws that are absolutely given. Everything can be calculated in advance and can be described in mathematical formulas and algorithms. There are no unforeseen events that cannot be explained by cause or chance. Just as machines have no goal or purpose, neither does the universe. Evolution in the universe is just blind chance. There is no goal or purpose in biological evolution. It is only about reproduction and the transfer of genes to the next generation. The question is, if everything in the universe consists of matter, how can we then explain that unconscious matter can become conscious through humans? For some unknown reason, the universe is waking up and becoming conscious of itself through humans.

If the universe is of unconscious and mechanical nature, then humans would also be of an unconscious and an mechanical nature. Everything we think and do would be of a mechanical, unconscious nature without goal and purpose. All actions would be unconscious, and we would be like robots and machines. It is to deny human free will to explain away consciousness. If humans do not have free will, and cannot make choices, then there would be no possibility for change and development. Research is based on our ability to make choices, make decisions, and respond to evidence. It is our ability to experience and reflect on our experiences that distinguishes us from a machine.

According to the Australian philosophy professor David Chalmers, there are three different ways to explain consciousness in relation to physicalism. One says that consciousness can be explained by physicalism when we have achieved sufficient knowledge. The other says that physicalism is the only thing that can explain the ultimate reality. Consciousness is an secondary epiphenomenon in the brain and a consequence of the brain. According to this view, there is a causal relationship, but it only goes in one direction. Consciousness does not affect physical processes, but physical processes affect consciousness. The third says that consciousness and physicalism are separate conceptual explanations of the ultimate reality, but there are causal relationships in both directions. Then consciousness affects physical processes, and physical processes affect consciousness.

Physicalism describes behavior, structures, and functions but has no explanation for consciousness. The question of why there is a first-person experience of the physical world remains.

Some physicists argue that there is a relationship between consciousness and quantum physics. In all quantum physical experiments, the state of a system seems to be defined by the information that can be made available to the observer. It is an interpretation that is highly debated. It is possible to calculate probabilities for what an observation of the system might yield in a measurement, but it can only give probabilities and says nothing about the observer. It can say everything about the state of the system, but cannot predict what happens during the measurement. According to quantum physics, the ultimate foundation of reality is a wave function that collapses upon measurement, but it has no explanation for why it does so. All we know is that it looks like a collapse from the perspective of a conscious observer during the measurement.

Chalmers asks a question of what happens during the measurement. Since it cannot be explained, it leaves room for consciousness. There is a causal opening in quantum physics perfectly suited for consciousness to fill. It provides a perfect "loophole" for the influence of consciousness on the collapse of the wave function. If we cannot explain consciousness in terms of spacetime, mass, and energy, we must acknowledge consciousness as the ultimate foundation of reality. This overturns the closed system in physics, which describes space, time, and mass.

The idea of consciousness as the foundation of reality was highlighted by philosophers in the late 18th century and early 19th century in a movement called German Idealism. At that time, German Idealism was a strong movement associated with religion and a view of nature as something transcendent or divine. In the late 19th century, natural scientists and intellectuals began to advocate an evolutionary theory launched by the British biologist and geologist Charles Darwin, that the theory of nature was an explanation for social conditions in society. Nature was a battleground where species fought for living space. By emphasizing the struggle and competition between humans, this view of nature could be applied to human society. In the 21st century, idealism reappeared through theories such as panpsychism and cosmopsychism.

Bernardo Kastrup

Bernardo Kastrup is a Dutch computer engineer and philosopher known for his theory of analytical idealism. In his book "*The Idea of the World*" (2019), he presents his theory of analytical idealism. Many interpret the idea of German idealism—that reality is fundamentally mental—as pertaining only to individual consciousness. According to Kastrup, it is not just individual consciousness that is mental, but all life in universe is mental, and everything can be reduced to universal consciousness. The term analytical idealism was chosen to indicate that he uses an analytical method to explain consciousness as the foundation of reality. Kastrup holds doctorates in both computer engineering and philosophy, and he is known for having published many theoretical reflections on the mind-body problem.

Kastrup asked the question whether it really was possible to simulate within a computer the flow of information that was present in a human mind. In his attempts to simulate conscious AI systems, he began to doubt that it would be possible to create conscious agents that could feel and experience things. There was no reason to believe that a simulation would suddenly be conscious or feel anything. He began to wonder if it was really possible to create consciousness from unconscious matter. He had never considered what matter actually was and questioned the assumptions that physicalism rested on.

Physicalism is based on the assumption that everything is matter, produced by our brain, and only studied how nature behaves. Computers could not provide qualitative experiences. They could only change structures and functions. He realized that it was not possible to draw deductive conclusions about qualitative experiences from physical information systems. It is possible to simulate information, but simulation is not the same as feeling qualities.

Physical parameters can´t be transformed into qualities or interpreted as qualitative experiences, as they are merely abstract entities. We can do whatever we want with artificial systems, but there is no reason to believe it would be the same as qualitative experiences. This led him to realize that the physical paradigm under which he had worked for so many years was no longer sustainable for qualitative experiences. Physicalism assumes that matter is the ultimate foundation of reality and that the brain produces mental experiences.

Kastrup argues that universal consciousness is the ultimate foundation of reality and that the brain is the external manifestation of our mental processes. If everything is mental, then matter is also an aspect of universal consciousness. All matter can be seen as representations of external mental processes that appear on our screen of perceptions. Physicalism cannot solve the hard problem of consciousness. If we start from the universal consciousness as the foundation of the ultimate reality, the hard problem of consciousness is solved.

Kastrup argues scientifically for his theory. He uses logical reasoning and empirical evidence, not relying on preconceptions or metaphysical beliefs. Normally, natural science is associated with the use of logical reasoning and empirical evidence, but it can just as well be used for idealism, which is normally associated with metaphysics and introspection. Evan the idea that the ultimate reality is mental can be understood through logical reasoning and the development of empirical evidence. Analytical idealism provides a meaningful scientific explanation and understanding of the world. It doesn´t put us in a dilemma to solve the hard problem of consciousness that physicalism does, says Kastrup.

All our experiences are mental and part of universal consciousness. Our experiences are the only things we have direct knowledge of. Everything else is theoretical abstractions, which we have no direct experience of. According to Kastrup, there are two perspectives on our experiences. One is an internal subjective perspective, called a first-person perspective. The other is an external objective perspective, called second-person- perspective. Viewed from a first-person perspective, other people's consciousness contents are objective but subjective from within their perspective. Viewed from internal perspective, nature is objective but subjective from within nature's own perspective. There is an objective world that presents itself for our observations, which we call the physical world. It is objective in the sense that nature and its processes do not depend on what we think and feel about it.

Kastrup equates the mind with the psyche. The psyche is a field of pure subjectivity and potential for experience. It is only potential for experience, but when it is excited, there are experiences. Each excitation of the field is an experience of the subjective field. The experiences are different patterns of excitation, which can give rise to an infinite complexity. The psyche is potential for excitation, in the same way as a quantum field. When the quantum field is excited, particles appear as patterns of excitation. Without excitation, there is nothing but the potential for excitation. The human psyche is an aspect of universal consciousness. There is only universal consciousness. Universal consciousness has always existed, and everything that exists develops within it.

According to Kastrup, humans, as well as all other living organisms, are dissociated modifications of the universal consciousness. The living organisms that we share the world with are external appearances of other dissociated modifications. Through a process of dissociation, the field of pure subjectivity becomes fragmented.The universal consciousness undergoes a dissociative process in a natural way. It begins as an undissociated field of pure subjectivity before life arises. There is no dissociative aspect of universal consciousness before life. What we call life, biological metabolism, is external manifestation of a dissociative process in the universal consciousness.

The question of why dissociation occurre is analogous to the question of why life arises. It is a misplaced question because everything just happens in nature. Nature has a disposition for patterns of being and behavior and will behave according to regularities, which we call the laws of physics. Similarly, the human mind has its own disposition for patterns of being and behavior. Each mind has its own preferences, memories, and goals, but we are all part of the same universal consciousness. What we lose in dissociation from the universal consciousness is that we can´t do cognitive association of our thoughts with something happening beyond our personal self. The human mind is a dissociative disorder, says Kastrup.

In psychiatry, dissociative identity disorder (DID) is a term for that parts of the personality are separated in different subpersonalities, that have different voices, memories, experiences, and behaviors, which the person shifts between. This leads to a split personality, disturbs the sense of self, and feels as though someone else is taking control of the body. The normal interaction between memory, perception, and motor skills deteriorates, and dissociative identities form, each with its own sense of identity, with different thoughts, feelings, desires, and opinions. Each experiences itself as separate from the others. That they are unaware of each other can be seen when the person cannot awaken another subpersonality.

Dissociation can be objectively diagnosed. There are neurological imaging techniques for diagnosis making it possible to see dissociation and the fragmentation of the mind into multiple personalities in the brain. Studies have shown very different brain activity for people with the diagnosis of DID when compared to people who have simulated DID. In one study, a woman was examined with a dissociative identitie that was blind. When the dissociative identity took control of her body, no normal activity was recorded in the visual cortex in the posterior part of the brain.

In his book "*Why Materialism Is Baloney*" (2014), Kastrup writes that physicalism is less coherent, less empirically adequate, and has less explanatory value. It creates more problems than it solves and can be reduced to absurdity. It claims that consciousness arises from matter but has no explanation for how qualitative experiences can arise from matter. Matter is defined through numbers that are purely quantitative and have no quality in themselves. It is a world based on empirical observations that we cannot have direct access to. Our knowledge of the world doesn´t begin with matter but with perceptions. Perception is replaced with an abstraction called matter, and we believe that it is the only reality we have access to. Consciousness and mental activity are the only realities we have direct access to.

Physicalism is based on the assumption that the material world is the one that appears in our perceptions, but it is only external manifestation of internal mental processes. Everything that appears on the screen of our perceptions is representations of the physical material world. Matter is just what mental processes look like when observed from the outside. The correlation between measurable effects of experiences and brain activity is an undeniable fact, but it does not necessarily mean that the brain causes experiences. Both the physical world and the world of experience are real, but each in its own way. Physicalism makes the mistake of confusing descriptions with the described. It is a mistake akin to taking a map for being the territory.

Physics is a science of perceptions and behavior based on the idea that there is an objective world outside. What physics sees are quantitative descriptions of something that cannot be visualized. It can only study how nature behaves, but cannot explain what nature is. Physics can formulate mathematical theories to describe and predict the world, but it overlooks subjective reality. Mathematical models are seen as more real than the subjective world, missing the difference between the subjective and the mathematical world. Physics creates models and predicts behavior by using measuring instruments. Everything is about observations and measurements. There are no properties before the measurements are made and observed. What we see are only measurements, not the world itself. The world arises from something else and is not made of things that can be measured. Our senses are not transparent windows to the world as it really is, because we can only see what is consistent with our perceptions.

Kastrup makes an analogy with a pilot sitting in the cockpit of an airplane. All the pilot can see of the physical world is the instrument panel, and reality exists beyond it. It is a world that can be measured, but at the same time, there can be a full storm in the reality outside. If the physical world is just the dynamics in the measurement windows, then it is only the behavior of the measurement needles in the measurement windows that physics studies. The evolution has equipped humans with an instrument panel instead of a direct window to reality so that we can survive and not be overwhelmed by all the information that floods us. The panel with all these instruments does not measure anything, and does not transfer any information other than about itself.

Problem arise when we cling to the story that physicalism provides even though it only tells us what the instrument panel says. It is a worldview that can lead to nihilism and a sense of meaninglessness. Our life is not just about surviving and passing on genes but also about finding meaning. The goal is to work for the development of life, become conscious beings, and find our place in the universe. The universe becomes aware of itself through humans. The question of why we exist and are conscious beings is more fundamental.

Physicalism is the main paradigm we live and function in. It says that there is an objective world, but we do not see the world as it really is, only what is useful for survival. It says that consciousness is a product of matter and that consciousness is generated by physical arrangements of matter. To find an alternative, we need to look at what explains and does not explain materialism.

One way to explain something is in terms of something else. For example, a body consists of organ systems, which in turn consist of tissues, which consist of cells, which consist of molecules, etc. Eventually, we reach a level that cannot be explained in terms of something else. This level is called ontological primitive, and it is the only and ultimate reality that gives rise to everything.

String theory and quantum theory explain the ultimate reality in two ways. String theory teaches us that the ontological primitive is vibrating superstrings that give rise to everything, but it provides no explanation of what superstrings are. Quantum theory teaches us that the ontological primitive is a quantum vacuum. It is a quantum vacuum that gives rise to everything, but it provides no explanation of what a quantum vacuum is. No one has seen a superstring or a quantum vacuum in reality. We can assume that universal consciousness is the ontological primitive instead of a concept of thoughts. It says that everything that exists is excitations in a universal consciousness.

In his book "*Decoding Jung's Metaphysics*" (2021), Kastrup gives his interpretation of the Swiss psychiatrist Carl Gustav Jung's theory of the psyche. According to Kastrup, Jung views matter from an idealist existential perspective. Jung distinguishes between the collective unconscious and the personal conscious. Kastrup interprets the collective unconscious as synonymous with phenomenological consciousness. It consists of psychic material loosely associated with each other, changeable, dynamic, and fluid. The psyche is fundamental, and not derived from anything else. It is the psyche that creates our perception of the physical world, how it appears when observed through our sense organs.

Kastrup interprets the personal conscious as a synonym to metacognitive consciousness. It is a higher level of mental function, organization, and structure than the level of the collective consciousness. It is a weave of cognitive associations held together by the ego in a more or less structured and organized way. Metacognition is not just experience but includes our ability to know that we have experience. It represents reflection and control over mental activity.

The capacity for metacognitive consciousness is not built into the universe from the beginning. Universe operates according to certain laws but becomes more aware of itself through the human capacity for metacognitive consciousness. We humans belong to a group in nature that is not only instinctive but also has the ability for metacognition. Without metacognition, we would be trapped in an instinctive world. Metacognition frees us from an instinctive life. It allows us to bring order to our minds and find a place for our experiences; otherwise, we become victims of our experiences. We channel a potential to reach a higher function of consciousness through metacognition. We can never return to a purely instinctive life, says Kastrup.

Federico Faggin

Federico Faggin is an Italian physicist, computer engineer, and inventor, known for his contributions to the development of the first commercial microprocessor and as a co-founder of Zilog and Synaptics.

Along with the Italian quantum physicist and professor of theoretical physics Giacomo Mauro D'Ariano, Faggin has developed a revolutionary theory of consciousness as a fundamental aspect of the nature of reality. This theory challenges the conventional views of classical physics, quantum physics, and free will. The classical physics describes reality in terms of particles that move in orbits, interact, and follow the laws of physics in spacetime. It is deterministic and reductionist, and it is not compatible with the concept of free will. In classical physics, the behavior of elementary particles and our actions are governed by impersonal laws, regardless of our thoughts and feelings about them. We are made of atoms and molecules and controlled by impersonal laws that dictate their behavior.

Quantum physics reveals a completely different view of reality. It describes reality as a unified field where particles are excited states of the field and do not exist as independent, separate entities. The quantum view of reality is holistic and compatible with free will, although it does not claim that free will exists. In quantum physics, there is superposition and entanglement, where new entanglements create wholes that are more than the sum of their parts. Everything is interconnected with everything else. This is a unique property of quantum physics that distinguishes it from classical physics. The quantum world is an infinitely larger world than the classical world of spacetime we live in. The quantum world and the classical world exist on the same continuum, ranging from infinite possibilities to just two possibilities. Quantum physics is more fundamental than classical physics. Classical physics is a special case of quantum physics. They can coexist in parallel and are both valid.

The theory posits that consciousness is a property of a quantum system in a pure state. A quantum system in a pure state is a coherent state. It cannot be known from the outside but can only be known through the system that is in that state. It is the representation of the inner space of our experience and is what we feel because we have all experiences within ourselves. It is private, just like our experiences are private. It is not made of parts but is an integrated experience as a unit and evolves as a unit. Consciousness is reflected in the pure state of the quantum system.

According to this theory, quantum physics is a gateway to a new paradigm that starts with consciousness as primary, where quantum physics and classical physics are special boundary cases. This means that the new theory provides a broader background for understanding reality. The theory can explain many other phenomena that have not been explained by quantum physics and classical physics. Quantum physics will still be valid, but it will be a special case of a broader theory, in the same way as classical physics is a special case of quantum physics. The theory requires a rethinking of several assumptions that physics rests on.

Quantum physics has given us a picture of reality that is far from our usual experience. Much of what happens in quantum physics is strange. Consciousness can provide an explanation for these oddities in quantum physics. Science has underestimated consciousness because it is so natural. He encourages science to use consciousness as an instrument for scientific investigation. There is enough evidence for science to leave the view of consciousness as a byproduct of the brain.

Faggin describes in the autobiographical book "*Silicon: From the Invention of the Microprocessor to The New Science of Consciousness*" (2021), his motivation for exploring consciousness. While working on creating conscious machines, he questioned whether it was really possible to make a machine conscious. Consciousness could not arise from a computer that only had electrical signals. His strong desire to understand consciousness led to an extraordinary experience. He felt himself to be one with the entire universe and felt a great love. It was an experience entirely different from the feeling of being separate from the world.

After this experience, it became clear to Faggin that his consciousness must be a much broader aspect of reality than his thoughts. Consciousness must be the starting point for reality as opposed to the endpoint in evolution. It was far beyond what he had previously thought and believed. He wished to understand this experience, which was not the same as intellectually acquired knowledge. It was an experiential knowledge more powerful than any intellectual knowledge. This experience revealed a reality he was not prepared to talk about. It led him to break with physicalism, which he had relied on for so long.

Faggin decided to explore consciousness introspectively. He gradually became aware of how many psychological blockages he had, and realized how much he wanted to believe in certain ideas to protect his need for control. Understanding himself was essential for freeing his ability to invent things. It took twenty years before he was fully convinced that the physicalist view of consciousness must be wrong. It only says that consciousness is a material phenomenon but provides no explanation for what matter is or how consciousness can be derived from matter.

Consciousness cannot be explained by fundamental entities that lack consciousness. Consciousness must be fundamental, irreducible, and be an original substance from which matter emerges. This substance must have the capacity for self-reflection, allowing it to know itself. Perception of space, time, matter, and energy can then emerge from something that has the capacity to know itself. By starting from the hypothesis that consciousness has the capacity for self-reflection, that is inherent in the substance from which everything is made, it becomes easier to explain how space, time, matter, and energy can emerge from something with consciousness rather than having to explain how consciousness emerges from inert matter.

Physicalism says that consciousness is a byproduct of the brain and that consciousness is not real in the sense that it has no causal power. Consciousness must be a primary aspect of nature. He studied biochemistry and neuroscience, but to his surprise, he found that the word consciousness was not mentioned in the books he read.

These books only described how the brain worked with biochemical or electrical signals. It seemed as if it was assumed that consciousness was identical to electrical signals. Consciousness must be more than blindly reacting to sensory signals. Faggin understood that it was impossible for sensory experience, feelings, and thoughts to arise from the brain. To perceive through perceptions, feelings, and thoughts is a completely different class of phenomena.

Many physicists have argued the possibility to create conscious computers. They believe that if computers only consist of electrical signals and the brain only consists of biochemical and electrical signals, then it is possible to create conscious computers. It is a view of humans as machines with only the functions that machines have. A computer can never become aware of itself or know what it is doing. It only does what it is programmed to do.

He disagrees with the physicists who claim that the universe and humans are machines without purpose or meaning. This view is based on the assumption that matter is a fundamental and irreducible property of nature. It is an unproven premise that cannot explain the existence of the consciousness we experience internally. We are conscious beings acting in a physical world where our bodies are part of the physical world. At the same time, we live in an inner reality consisting of sensations, feelings, and thoughts. It is through our representations, based on the processing of information through our senses and brain, that we come to know the world. If everything had been made of matter and lacked consciousness, it would be impossible to have an inner experience. Consciousness must be more than the energy that matter, space, and time are made of. Reality is more related to information than matter. There is no evidence that matter is the substance of reality because reality is linked to quantum information.

Quantum information is a representation of experience, not a representation of information. It does not describe information because that would make it possible to copy and share it. Quantum information cannot be copied and cannot be shared in the same way that a computer can. It cannot be copied because it is private, similar to the contents of our consciousness. Quantum information is not the same as classical information.

Quantum computers work with qubits, while classical computers work with bits, and a qubit carries more information than a bit. A qubit represents an infinite number of states, while a bit represents two states. A bit in a classical computer can be copied, observed, and read without disturbance, while a qubit cannot be copied, observed, or read. When a qubit is observed, it is changed to either zero or one, with a certain probability. The outcome cannot be predicted, only measured as a state, and measurement collapses the wave function. The collapse of the wave function is how the quantum state becomes a physical state in spacetime. It is a step from the quantum state, with its infinitely many possible states, to a state in the physical world. The quantum computers can operate with different states att the same time due to entanglement. In both classical computers and quantum computers, there is no independent whole. They can only create a whole as the sum of parts, and it is algorithms that interact to form the sum of the whole.

Consciousness cannot be a property of a computer. If it were a property of a computer, it would be possible to create identical copies of ourselves in thousands of computers. But this cannot be possible because it contradicts the fact that experiences are private and cannot be copied. Humans have qualities that artificial intelligence (AI) and robots cannot possess. AI can complement human intelligence, consciousness, and the free will. AI is about transforming a symbol into another symbol without meaning. Through a process of meaning, as opposed to symbol transformation, we can make more intelligent decisions than a computer can.

True intelligence requires understanding and meaning. Computers have no understanding or sense of meaning. It is we humans who give understanding and meaning to information, not computers expressed in algorithms. Humans gain their meaning through the experience of information that comes from our senses. Computers carry information, but it is through consciousness that information is translated into a subjective experience. Computers have no free will because free will is associated with meaning. Freedom is not in the syntactic rules of the symbols used to represent meaning. If we believe that symbols are fundamental, then the meaning of our lives and existence is lost.

Physics focuses only on a symbolic external reality and denies that the meaning and purpose we feel can have a direct impact on reality. It is humacs who define the meaning and purpose of life. Physicists doen´t consider consciousness a factor to be reckoned with when they investigate reality. They assume that consciousness is a byproduct of the brain but does not provide an adequate explanation of how matter can produce consciousness. It can be valuable for physics to explore consciousness. Mental and emotional aspects of reality are also valuable to explore.

Physics does not recognize the existence of an inner aspect, but our experience has both an inner and an outer aspect of reality. The inner aspect represents what is subjective, experiential, and semantic. It is thoughts, feelings, and sensory impressions that give meaning to experiences. It is an experience that is private. The outer aspect represents what is symbolic. These are symbols we use to communicate with each other and experiences we can share with others. Created symbols do not have the same level of reality, even though they are part of the experience.

The physical universe is the symbolic manifestation of the semantic inner reality. These are separate, but they reflect each other in some form. The development of the physical universe reflects the semantic development of the individual. One reflects and supports the other. Both arise and develop simultaneously, and matter is created so that consciousness can reflect on itself. Development of matter is a mirror for consciousness to know itself.

The universe is created so that the conscious self can make its own experiences and learn about itself through reflection. The desire to know oneself is the most significant aspect of consciousness. This is a proposal that he explains with the metaphor that matter is the ink that consciousness uses to learn about itself. Similarly, we can use writing as a way to learn about ourselves. We can use writing to reflect on the words we write and on our thoughts. We need to write to reflect, think better, and see if our logic is correct.

Faggin defines consciousness as the ability of a self to perceive, feel, and experience itself and the world. Besides feeling the outer world, consciousness can also be directed inward to feel itself. This self is experienced as an independent agent with free will. It has the ability to make decisions and act based on understanding and be consistent with intentions and goals. The self makes a free choice based on understanding, intention, and purpose. Freedom is the idea that each self can do whatever it wants instead of being forced by another agent to go against its will. Intention and purpose are integral parts of free will and cannot be separated from it.

Artificial intelligence (AI) can be very useful with the goal of creating a better society by doing what AI is good at but what we cannot do. Physics has given us great technological inventions, but if we do not understand ourselves, our technology can be used against us. It is only within ourselves that we can understand what is right and wrong. We can create a better society through the creative intelligence and free will that only exist in consciousness.

The mathematical perspective, which says there is only an objective world made of matter, energy, space, and time, is deeply flawed. Mathematics cannot capture the essence of what consciousness is, but the interaction between consciousness and material objects can be described with mathematics. Mathematics can describe properties manifested through matter. Physical laws are expressions of mathematics. We need a mathematical theory that is primarily based on consciousness. If consciousness is primary, then new predictions can be made that can be scientifically verified. It can provide us with testable predictions and help us understand how consciousness can be used as a method for investigation

19. PHENOMENOLOGY AND EXISTENTIALISM

Phenomenology is a theory in philosophy that particularly studies the relationship between perception and objects we perceive, and does´t primarily concern itself with causal relationships between objects. Phenomenology focuses more on the content of consciousness and aims to explain or describe phenomena as they appear to us. It places less emphasis on linguistic and logical analyses, as in analytical philosophy, and more on the direct understanding of human existence and experience. All our knowledge is mediated through experiences, and we perceive the world of phenomena through our senses. These are our experiences of everyday unreflected life, and our world of memories and future expectations.

The basic idea in phenomenology is that consciousness is always directed towards some object under some aspect. All mental content is directed towards objects. Our directed experience concerns not only material objects but all kinds of objects, such as music, mathematics, not to mention our thoughts, feelings, and memories. The focus is on phenomena, whether they come from the external world, a memory, or a dream. Whether the phenomenon is fantasy or real is irrelevant. There must be something in the mind that explains this direction towards an object. What accounts for direction towards an object is called ”intentionality”, whih means a conscious orientation towards an object. The intentional act is a creative act of meaning. Every intentional content is a kind of description of reality. We do not act based on an objective reality but based on our understanding of the meaning of the situation, and form an understanding of the situation's meaning. This is what Husserl calls the meaning-creating act. Consciousness is the basis for understanding everything.

Edmund Husserl

Edmund Husserl (1859-1938) is a German philosopher and professor of mathematics, known as the father of modern philosophical phenomenology. Husserl argues that we can never know if objects in our consciousness have a separate existence independent of our experience of them. We cannot capture a description of the truly "objective" but only our experience. There is no need for us to make any assumptions about their independent existence. We have a direct access to experience as objects in consciousness and can examine them as such. It is impossible to separate states of consciousness and objects of consciousness. As soon as we start analyzing consciousness, we find that consciousness is always conscious of something.

Martin Heidegger

Martin Heidegger (1889-1976) is a German professor of philosophy, a student, and follower of Husserl, known for his contributions to phenomenology and existentialism. According to Heidegger, all philosophy from Plato onwards has been thinking in incorrect categories. The central task is to investigate and understand humanity's place in the world. In his book "Being and Time" (1927), Heidegger poses the question of "being." For Heidegger, "being" means "being there," in contrast to science, which is about "knowing." Being means complete presence and being there. It is not through theories and analytically rational thinking that we find meaning in life. Science cannot show humans the meaning of their life. We have forgotten to marvel at our being. The question of being appears as empty nonsense, but we must ask why there is being and not nothingness, says Heidegger.

The primary concern is not scientific understanding but understanding the nature of being. It is impossible to provide an objective account of consciousness because it is intertwined with the world and being. We are not primarily observing and knowledgeable beings but are already involved in the world from the start. We do not start from a theory of the world; we start from how we should handle the world. It is impossible for humans to understand the world by abstracting themselves from it and trying to understand it through a theory because the human situation is to be in the world. Most activities are not determined by conscious choices and states. It is only when something goes wrong that we become a problemsolver and a rational thinker. For example, we can use a hammer without needing to analyze, think, or pay attention to it. It is when something goes wrong with the hammer that we need to figure out what the problem is. Then we begin to see the hammer as a problem to solve, and need to think rationally.

Heidegger sees existence as potential for possibilities. Freedom is to be open to new possibilities; that anything can happen at any moment. The present moment has meaning because it opens up a possible future. We have the task of becoming aware of ourselves and creating ourselves when we face an unknown future. Faced with our mortality, the question of life's meaning naturally arises. The problem of existence is not that we live in anxiety about being free. It is when we oppose our freedom and do not understand that we do so that existence becomes a problem. Heidegger is one of those who foreshadowed existentialism.

Existentialism

Existentialism is a philosophy oriented towards exploring questions related to the meaning, purpose, and value in the human existence. Common concepts in existential thinking include existential crisis, dread, and anxiety in the face of an absurd world, but also concepts such as authenticity, courage, and virtues. It emphasizes human existence and explores subjects related to humans' experience of meaning, purpose, and value. We are thinking, feeling, and acting beings who are free to act and make choices. Questions of responsibility, ethics, and morality are central.

Existentialism is a movement that emerged in literature and philosophy in the late 1800s and early 1900s and began in Germany in the period following World War I. The cultural atmosphere was permeated by post-war existentialism. There was a sense of disorientation in the face of what seemed to be a meaningless world. The most extreme view was nihilism, the belief that there is no meaning at all. The theme for literature in the 1900s was that human existence is characterized by alienation, emptiness, and boredom.

Jean-Paul Sartre

Jean-Paul Sartre (1905-1980) is a French writer, philosopher, and Nobel laureate in literature in 1964, and perhaps the one most associated with existentialism in the post-war period. Sartre played a significant role in spreading existential ideas to a broader audience beyond academic circles. What is now called "existentialism" is the direction developed by Sartre and a group around him. Sartre's existentialism emphasizes the individual freedom, responsibility, and the creation of meaning in a meaningless world. According to Sartre, we are thrown into a chaotic world, projected onto a world and reality, in which we must create meaning in our existence. Initially, there is no meaning, but as unique individuals, we can make choices and actions in life that create meaning. Sartre argues that we have total freedom, and nothing can prevent us from making free choices at any moment.

Every choice reveals what we think and how we should be. This means that we must always act as if everyone can see us because it provides a picture of what we believe in and what we should be as humans. Reality is a choice and a freedom we have, but freedom also brings anxiety about making wrong choices. We are condemned to be free, and the impermanence of existence brings uncertainty and anxiety. According to Sartre, there is no design for how humans should be. There is nothing in our nature that dictates how we should live. We are what we make of ourselves and define ourselves through our actions. The choices we make have no predetermined values. Through our choices and actions, we choose our own morality, but this becomes significant only after action or inaction.

20. HUMAISM AND LIBERALISM

Humanism is a cultural movement based on a worldview that emphasizes individual value. It is an ideology that originates from the individual's free will, freedom, and the right to determine meaning. It asserts that the individual's free will is the highest authority of all and is based on the assumption that there is a free, authentic self. It arose as a response to the experience of meaninglessness when religious narratives could no longer provide meaning to existence. Humanism asserts that humans should be the foundation of meaning, not religions. It is human feelings that are the source of all meaning. All humans have feelings, and all feelings are equally valid and cannot be questioned. The ethical rule is that if something feels good, then it is right. It is from within ourselves that we derive meaning based on knowledge and experience. Life is a gradual process of inner changes leading from ignorance to enlightenment through experiences. The ultimate goal is to fully develop knowledge through various intellectual and physical experiences.

Liberal humanism is a political ideology based on the same ideas of free will, freedom, and the right to determine meaning. Liberal humanism teaches that the individual trumps the collective. It is the individual who should determine what opinions and values should prevail, not the collective. The individual has authority because only the individual has access to the inner self and knows themselves. Change is achieved through self-exploration, not by collective institutions.

Social liberalism is a political ideology based on ideas that emphasize the group's right to determine meaning.

Social liberalism teaches that the collective trumps the individual. It is the collective that should determine what opinions and values should prevail, not the individual. The collective has authority because only the collective has access to more than what the individual's inner self has. Change is not achieved through self-exploration but by collective institutions. Social liberalism criticizes liberalism for placing too much emphasis on feelings and not paying attention to how actions affect other people. It is important to see who owns power and production, and only collectively can systems be changed.

Power is about controlling people to get tasks done. Those who have the power to determine what opinions and values should prevail can control how we think and act. An important value for humans is to have freedom to think, act, and live according to what feels true. This choice cannot be made at a collective level. It is only at the individual level that this value can be realized.

In society, our institutions depend on economic resources to gain the power to fulfill their role as institutions. They survive only by demonstrating their power to make a change in society that is perceived as valuable to obtain resources for their existence. Even scientific institutions depend on economic resources to gain the power to fulfill their role as institutions. They survive by demonstrating their power to make a change in society that should be perceived as valuable to obtain resources for their existence. All institutions prioritize power over truth and are bound by conventions.

People often seek identity and security by identifying with an ideology. Ideologies show us how to think and feel. It gives a sense of belonging and collective consensus of opinion, but it does not guarantee the freedom to live according to what feels true. There is nothing wrong with having an ideology, but it becomes a problem when we identify with an ideology. It leads to a situation where we experience it as a catastrophe if something happens that goes against what we believed to be true. When we realize that what we believed in can no longer explain what we previously thought to be true, we are shaken to our core and no longer know who we are. We have lost the narrative that gave us our identity and security. The risk is getting stuck in a collective consensus and performing actions that cause unnecessary suffering for ourselves and others.

It is sufficient for a consensus of opinion to establish rules for morality and ethics. When we identify with moral opinions and rules, there is no need to consider the consequences of our actions for ourselves and others. Morality and ethics can guide our actions and decisions, but they do not guarantee that they are right. Moral duties are an expression of good will, but they are no guarantee of good morals. Morality deals with rules and principles that individuals believe it is their duty to live by.

Moral rules are influenced by traditions and vary among different cultures and individuals. Ethics answers the question of what constitutes a morally good life. Ethical studies deal with different moral views and the moral language. They can be related to the consequences of actions or absolute principles regardless of consequence. Demanding ethics and moral governed by ideology is no solution to human problems. Collective solutions are tested through visions of both fascism and communism.

2. THINKING AND BEIING

A fundamental question is the relationship between thinking and being. Thinking and being offer two different perspectives on the experience of the world. Through thinking, we experience the world as consisting of separate abstract forms. It is a world that is limited and bound by forms and time. Through being, we experience the world as a continuous flow of sensory perceptions. It is a world that is unlimited and unbound by forms and time.

It is well-known in psychology that we always distinguish something as a figure and perceive the rest of the perceptual field as the background. This is especially evident in experiments with reversible figures. The lines and surfaces that reversible figures consist of can just as easily represent one figure as the other. It is impossible to see both figures simultaneously, but we can shift between the different ways of seeing. How a reversible figure is perceived depends on what is foreground and what is background. If one figure is in the foreground, a different image emerges than when the other figure is in the foreground. In the same way, thinking and being are different perspectives on the world and depend on what we perceive to be the foreground and background. If thinking is in the foreground, a different picture of the world arises than if being is in the foreground. Being is overshadowed by thinking. Our thinking is important, but when thinking occupies all our attention, an imbalance is created. Our entire world then comes from our mental activity, and there is no room for being. Thinking absorbs all our awareness, and we live in a world of abstractions.

Within us, there is a constant background noise of inner self-dialogues that we are often unaware of. It is a noise of involuntary, repetitive, and irrelevant thoughts that border on the unconscious. They distract us, and create unnecessary stress, tire us out, and steal vital life energy. These are self-related streams of thoughts that cause us to get stuck in loops. They are very difficult to stop and prevent us from being able to relax. When we are in the being, we are free from involuntary, unproductive and repetitive inner self-dialogues. We feel pure awareness, presence, and inner stillness. There is only pure being, a sense of life, and being alive. When we are no longer in the being, the mental noise of thoughts returns, and the stillness is lost.

Most of the time, we are busy thinking about what needs to be done and become absorbed in thinking and doing. When we lose ourselves in thinking and doing, we can always return to being. Many people are only aware of doing and unaware of being, which creates stress. Stress arises when we want to do many things simultaneously and try to accomplish as much as possible in the shortest time. We can do one thing at a time, as each step is equally important in achieving goals. This brings better quality to what we do, and we work more effectively when we are not occupied by thoughts of goals. We believe that only when we have achieved our goals can we relax and feel satisfied and happy. This feeling lasts only a short while, and then we are on the hunt for new goals. It is important to achieve goals, but when all our time is spent seeking relaxation and happiness, we miss that the goal of our quest for relaxation and happiness is found in being present.

We can be present when we do things and not perform them mechanically. We can have brief moments of presence in our daily lives without noticing it. To maintain full attention to what we do, we need to practice being present. Life is not just about thinking and doing but also about being. There must be a balance between doing and being. We need to slow down and allow ourselves to just be. In the being, there is no need to be anyone other than who we are. When we are aware of what happens hear and now, without judging what happens, we are in a state of pure being, called conscious presence.

Conscious presence is characterized by high alertness combined with a calmness that comes from the quieting of thoughts. It is being alert and simultaneously very still. Being alert is a heightened state of wakefulness, and the highest form of wakefulness. It means being vigilant and prepared for anything to happen. When thoughts come, we can always redirect our attention away from thinking to conscious presence. The redirecting attention away from thinking to conscious presence makes the body feel lighter and filled with energy. We can feel how life and energy pulse and flow in the body. When we redirect our attention from thinking to conscious presence, energy is released, and we become more aware. Awareness and the energy level increase analogously to a lamp shining brighter when a dimmer is turned up, increasing the current supply.

Just as it is not possible to force sleep, it is not possible to force conscious presence. Our inability to stop thinking is a sign that thinking becomes compulsive. This creates a dependence on thinking, becoming an obsession with constant thinking. When we rise above thinking, thoughts lose their hypnotic power. There is no problem with thinking, but not being aware of what is happening in the present can be a problem. In mindfulness, we can enjoy being because there is no longer any dependency. We can still do and create things, be active, explore, and learn new things, but there is no sense of compulsion or necessity.

Examples of activities performed in conscious presence are surfing, mountain climbing, and stand-up comedy. Techniques can be learned, but in the moment of the activity itself, thinking must cease to act from conscious presence. It is by letting go of thinking that these activities can flow spontaneously and freely. Surfing, climbing mountains, or any other physical activity temporarily frees us from thinking, and we become more aware and present. The deeper reason for the feeling of liberation comes from being in conscious presence..

22. MENTAL TRAINING AND MEDITATION

There is a rapidly growing development of techniques focused on mental training. Our understanding of the brain, and its plasticity, has increased the interest in using mental training to improve cognitive abilities. The brain has the ability to change and develop with increased training and experience. It can reorganize itself and improve when we learn new things. Cognitive abilities can be trained, and neural connections in the involved areas can be strengthened. They can be maintained and enhanced in the same way physical abilities can be improved through physical training. The brain is malleable and changes its structure depending on the surrounding environment and experiences in our lives.

There are two main types of mental training. One type consists of exercises focused on performance, thinking, and doing. The goal is to develop cognitive abilities to handle a specific task. These exercises put the body in an active and tense state, leading to increased activity in the stress system. Examples of methods to improve performance can include various forms of reinforcement of motivation and self-confidence. The other type consists of exercises focused on conscious presence. The goal is to be free from the demands of performance, thinking, and doing. These exercises put the body in a passive and relaxed state, leading to stress reduction. Examples of methods for conscious presence include different types of meditation.

Two main types of meditation are the so called focused attention exercises and mindfulness exercises. Focused attention exercises involve directing attention towards a specific object. It can be a simple word, a phrase, or the rhythmic sound of a mantra. The goal is to become aware of distractions in attention and to redirect the attention from distractions by refocusing on the object. Mindfulness involves focusing on experiences in the current moment. It encompasses both internal experiences, such as thoughts, feelings, or bodily sensations, and external experiences, such as visual impressions, smells, sounds, and more. The goal is to break reactive behavior patterns and be more observant and willing to face experiences without judging or interpreting them. Focused attention exercises and mindfulness exercises can both be used for stress reduction.

Stress is a natural reaction in our body. Stress has the function of providing energy and preparing for increased performance. It can give extra energy and be helpful in achieving more. When we become stressed, the heart starts beating faster, we breathe faster, muscles tense up, and blood flow increases to muscles and the brain. Stress can be triggered by physical and mental strains, known as stressors. Examples of stressors are situations where there are demands we perceive as exceeding our capacity. It depends on our self-perception in relation to our resources.

Chronic stress conditions have negative consequences. Our memory function, perception, and ego function deteriorate, and can lead to mental and physical health problems. For stress to have a positive effect, the body also needs to rest after exertion. Stress that is too high for too long leads to overexertion. Overexertion involves repeated activity in the sympathetic nervous system and leads to chronic stress conditions.

The sympathetic nervous system is activated when we need to make an effort and mobilize energy to act and move. Energy may be required to get up in the morning or to do physical exercises. The sympathetic nervous system increases respiratory rate, heart rate, blood pressure, and blood flow to the muscles in the body. Cortisol, adrenaline, and noradrenaline are different stress hormones that activate the sympathetic nervous system. The "fight or flight response" belongs to the sympathetic nervous system. It is a survival response characterized by fear and preparation to either fight or flee.

The parasympathetic nervous system is activated when we need to relax, rest, sleep, and digest food. It is a system that activates for cell regeneration and repair, and a strong immune system. It decreases respiratory rate, heart rate, and lowers blood pressure. Parasympathetic nervous system includes what is called the "freeze response", which is a survival response that involves becoming completely still and immobile. Respiratory rate, heart rate, blood pressure, and oxygen intake decrease. It is a kind of shock state that can happen in response to an unknown or unexpected stimulus.

The sympathetic and parasympathetic nervous systems belong to the autonomic nervous system. The autonomic nervous system consists of nerves that control functions in various organs of the body that are not directly under conscious control. The autonomic nervous system cannot be consciously influenced, but it can be affected through specific breathing techniques. The amount of oxygen we take in through breathing affects the amount of energy released in the body's cells. By having the right breathing technique, it is possible to gain control over several physiological processes.

Herbert Benson

The American physician Herbert Benson (1935-2022) conducted in the 1960s one of the first scientific studies on meditation. Benson assumed that meditation could play a role in reducing stress. Benson describes two main components of meditation: the use of a mental tool such as a simple word, phrase, or other activity to repeat to prevent thoughts from wandering, a passive attitude, a calm environment, and a comfortable position. In his study, the word ”one” was used which the participants were asked to silently repeat until the word was barely audible. The word ”one” was chosen because it would not immediately lead to thought associations.

Benson found that during meditation, metabolism quickly decreased in just a few minutes. During meditation, a so-called relaxation response occurred. This term was used to denote the body's ability to relax the muscles and internal organs. The relaxation response was associated with the activation of the parasympathetic nervous system and showed a decrease in respiratory rate and heart rate, and reduction in blood pressure, metabolism, and brain activity. In his book "*The Relaxation Response*" (1975, 2001), the response is described as ´an automatic reaction triggered by a meditation technique designed for people in the Western world´.

At the same time, American physicist and physiologist Robert Keith Wallace worked on a dissertation that included a meditation technique called transcendental meditation. Wallace describes transcendental meditation as a technique for passive attention to a mantra. Through passive and repeated focused attention on a mantra, a state that Wallace calls transcendental consciousness or pure consciousness is induced. It is described as a wakeful hypometabolic state, distinct from deep sleep, dream sleep, and ordinary wakefulness (Wallace 1970, 1971, 1972). In recent decades, quantities of articles on meditation, of which several have been published in scientific journals. This indicates a great interest in these types of meditation.

PART III

KNOWLEDGE

In this section, I will address various forms of knowledge, different theories of truth, ways to justify knowledge, ontology, metaphysics, knowledge development, education, the value of knowledge, self-knowledge from , Western and Eastern perspectives, mathematics and the empirical world, physics and mathematics, scientific paradigm shifts, relativity theory, quantum physics, and the concepts of time and causality.

23. VARIOUS FORMS OF KNOWLEDGE

Knowledge can be classified into different forms, such as theoretical knowledge (propositional knowledge) and practical knowledge (procedural knowledge). Theoretical knowledge involves the ability to understand, explain, and articulate how things are interconnected. Practical knowledge involves the ability to perform certain tasks and skills. It is common to refer knowledge to theoretical knowledge. One definition of knowledge is as a learned theoretical ability to understand, reproduce, and apply information. Ultimately knowledge is about how we create understanding of reality.

The word knowledge comes from the Greek word "logia," which means knowledge. In ancient Greece, the Greek philosopher Aristotle spoke of three forms of knowledge, which he called Episteme, Techne, and Phronesis. Episteme represented theoretical knowledge, which explained why something was a certain way. It was knowledge that was true, certain, and distinct from opinion (doxa). Techne represented practical knowledge, which enabled one to perform a craft or practice an art. It was knowledge as skills and abilities. Phronesis represented good judgment and understanding. It was knowledge as ability to make well-founded decisions.

The Swedish philosopher Bernt Gustavson interpretes these forms of knowledge from a modern perspective. In his book "*Knowledge Philosophy – Three Forms of Knowledge in Historical Ligh*t" (2000), he interpretes episteme as scientific knowledge, with its requirements for generalizability and measurability. Techne is practical knowledge, which can apply to technology, economics, or craftsmanship. Phronesis is the ability and readiness to orient oneself in the world and meet situations with attention, imagination, and ethical reflection (wisdom).

Philosophy of science and epistemology deal with different questions about knowledge. The philosophy of science addresses questions about the specialized knowledge in various scientific disciplines. It can be questions about what distinguishes science from non-science, what characterizes a scientific explanation, and if science provides reliable knowledge or is influenced by ideologies or cultures. It can also involve questions about whether different explanatory models are required for the natural sciences, which deal with natural phenomena, and humanities, which deal with human activity.

Epistemology addresses more general questions about what knowledge is and the value of knowledge. In the analysis of knowledge, there are three criteria for knowledge, called truth, belief, and good reasons. Epistemology often starts from a definition of knowledge as statements that are both true and convincing. In a shortened version, knowledge is said to be "justified true belief." The idea of knowledge as "justified true belief" was introduced by the Greek philosopher Plato.in ancient Greece According to Plato, knowledge was a subset of truth and belief.

It has been questioned whether justified true belief are necessary and sufficient conditions. It may not be enough to have a strong belief and conviction to claim that something is true. People can claim to know something they are convinced of and have good reasons to believe is true, but it may still be merely opinions, prejudices, or misunderstandings. Opinions, beliefs, and convictions are immediate and irrevocable mental states. There is a need for something that justifies the claim to be true and likely. All theories of knowledge must be based on some form of truth..

24. DIFFERENT THEORIES OF TRUTH

The oldest way to determine that something is true is through logical reasoning. There are two different types of reasoning: the deductive reasoning and the inductive reasoning. In deductive reasoning, the conclusion is a logical consequence of the premises, which are assumed to be true, but it don´t provides any guarantee that the conclusion is true if the premises are true. Premises may have been assumed to be true, but are never certain. The reasoning might have relied on incorrect premises, and therefore lead to incorrect conclusions. There is also a risk of missing important facts due to the limited number of premises.

In inductive reasoning, the conclusion is not a logical consequence of the premises, but based upon empirical experiences. Based upon a number of events, a probable conclusion can be induced. This is called generalization, which means that from one or more cases, conclusions are drawn that it applies to all other similar cases. It is based upon the implicit assumption that similar things have similar properties, and that individual cases are representative of all other similar cases. Inductive inferences can never be certain to one hundred percent. They can only have a higher or lower degree of probability depending on the sample. They can lead to overgeneralizations, and thus result in incorrect conclusions. For example, when we discover an edible mushroom, we believe that all similar mushrooms are edible, but one might be edible and the other poisonous.

In attempt to define what constitutes a true knowledge, there are different theorie. Thre theories of what defines true knowlede are those that are called correspondence theory, coherence theory, and pragmatic theory.

The correspondence theory, which is the oldest, posits that a statement is true if it corresponds to the facts or reality. According to this theory, truth is a matter of how accurately a statement reflects the external world. It rests on the assumption that there is a direct relationship between thoughts or language and the actual state of affairs in the world. Criticism of this theory is that it is often difficult to verify if statements perfectly match reality, as our interpretations and our perceptions can be flawed. The external physical world can hardly have a structure that resembles the structure of language. A theory can only provide a simplified picture of reality, and never a complete picture of reality. It can be useful and valid on one level, but not valid on another level.

The coherence theory suggests that truth is determined by the logical consistency and integration of beliefs or statements within a system. A statement is considered true if it coheres with a set of other statements or beliefs without contradictions. The theory emphasizes the interconnectedness of ideas and the importance of a coherent and harmonious belief system. A limitation is that different systems can be internally coherent yet mutually exclusive, leading to relativism.

This means that it is the system that determines what is true. A statement can be coherent in one system, but not coherent in another system. This results in different degrees of truth, depending on how well statements are logically connected with the system. The question arises whether the system must be free from all contradictions to speak of coherence. There are not many systems that can have that degree of coherence, where all statements actively support each other, and fit together like pieces in a puzzle.

The pragmatic theory suggests that truth is what works or has practical effects. According to this view, a statement is true if it proves to be useful or satisfactory in practical application. Truth is something that evolves and changes, dependent on the outcomes of actions and experiments. This theory emphasizes the practical consequences and utility of beliefs. Critics of the theory is that the approach potentially can overlook objective reality.

Neither logical reasoning nor any theory of truth can provide true justified knowledge. Logical reasoning only deals with the formal conditions for how we should think for something to be true, and a theory only deals with interpretations and explanations of how different facts are logically connected. Neither says anything about what justifies knowledge.

25. RATIONALISM EMPIRICISM KANTIANISM

Research on what justifies knowledge can be divided into three main areas, known as rationalism, empiricism, and Kantianism. Rationalism refers to a direction within philosophy that says all knowledge of the external world originates from rational thinking. Rational thinking is the basis for justified knowledge. The formulation of a theory is determined by rationally grounded facts. Criterion for truth is intellectual and deductive. Knowledge depends on our mental functions and processes. The only reliable is what can be supported by logic and rational thinking. Only by using our intellect is it possible to attain certain knowledge.

René Descartes

René Descartes (1596-1650) was a French mathematician, philosopher, and the founder of rationalism. He is known for his philosophical statement "I think, therefore I am". In this statement, he saw a safe knowledge that he couldn´t disbelieve. Descartes describes in his book "*Discourse on the Method*" (1637) his method of doubting his beliefs. Through logical thinking and analysis, erroneous beliefs should be discarded. Descartes believed in introspection, guided by clear definitions and logical thinking. Problems in the world are due to unclear definitions and a lack of logical thinking. Descartes celebrated the individual's right to think freely. Ideas should be based on individual reasoning and not governed by authorities and traditions. All knowledge can be found within ourselves. We only need a 'quiet room and a rational intellect.'

In the 1600s, Descartes' teachings created a systematic school within philosophy for the first time and had a profound influence on Western thinking in general and particularly on science. It was the rationalists who gave the period "The Age of Reason" its name and place in history. Various degrees of emphasis on Descartes' theory and method led to claims ranging from assertion that logical thinking took precedence over other ways of gaining knowledge, to the more extreme claim that logical thinking was the only way to obtain knowledge. The most extreme had such high confidence in rational thinking that empirical evidence was considered unnecessary to secure knowledge.

Rationalism came to be set in opposition to all ideologies of a religious or irreligious nature. To be called a rationalist was to hold a worldview that had no place for the supernatural. Since the Enlightenment, rationalism has implied a focus on secularism and rational choices. With Descartes, the relationship between the individual and the world came to be described in terms of a subject that experiences and perceives objects. The most central question became what the relationship between subject and object looks like, and how we as subjects can gain knowledge of these objects.

Empiricism

Empiricism refers to a direction within philosophy that says that sensory experience is the foundation of justified knowledge. All knowledge of the external world originates in our sensory experiences, and the formulation of theory is determined by empirical facts. The criterion for truth is sensory, and not intellectual and deductive. Knowledge depends on perceptual experience of the external world, and not on mental functions and processes. The only reliable is what is supported by experiences. It is only possible to achieve safe knowledge through sensory experiences.

David Hume

David Hume (1711-1776) was a Scottish philosopher and historian, known as one of the founders of empiricism. According to Hume, people reason based on beliefs rather than rational thinking. Not many beliefs are driven by rational investigations of facts. Rational thinking is rather used to support beliefs and opinions. They are considered true or false solely based on belief systems.

People's thinking and reasoning are more governed by emotions and passions than by rational analysis and logic. Emotions are always present, and we must accept them. It is not possible to use logical reasoning to suppress emotions. It is not through logical reasoning that people engage with the world.

A variant of empiricism is positivism. The word positivism refers to a philosophical direction that says experiences come through a combination of our perceptions and thought processes. Empirical evidence is required for something to be considered true. Knowledge, derived from experience and based on empirical evidence, is called empirical knowledge

Kantianism

The term Kantianism refers to a philosophical direction that says the real world is inaccessible to us. What we believe to be an objective world is nothing but the world as we subjectively experience it, and it is only that world we can know anything about. We would not experience the world as an orderly, rule-governed place if time, space, and causality were not functions of our perceptual faculties.

Immanuel Kant

Immanuel Kant (1724-1804) was a German philosopher and founder of Kantianism. Kant's theory of knowledge is an attempt to reconcile rationalism and empiricism. One of his most significant books is "*Critique of Pure Reason*", published in 1781. With this work, he aimed to lay a safe foundation for philosophy and conceptual investigations about what makes knowledge possible. In the book he presents a way to gain knowledge about the world that does not solely rely on science and logical thinking.

Kant proposes a third way to knowledge and opens up a completely new method of investigation. He seeks to determine the conditions for our knowledge, such as when we perceive objects or try to understand them as causes and effects. Kant states that it is we ourselves, with our intellect, who create the temporal and spatial order in our world of experience. This world is therefore more like "phenomena" than as we scientifically explore it. The knowledge we have about things and phenomena is a human creation. It is conditioned by ourselves, and therefore we lack the ability to grasp the true conditions of things.

All knowledge begins with experience, but it does not arise from experience. Everything we experience, whether it is sensory impressions, emotions, thoughts, or memories, is experienced through our body's organs of knowledge, our senses, our brain, and central nervous system. Our perceptions depend not only on an external world but on internal mental functions and processes. On the one hand, there is the human interpretation of reality, what we are capable of perceiving. On the other hand, there is an objective reality that is independent of us and our perception. There is no reason to believe that these two kinds of existence are the same.

There is a "filter" between humans and the objective external world, which makes us interpret reality but never fully experience it. This filter can be likened to a pair of colored glasses. In the same way blue-tinted glasses make reality appear blue-toned, we perceive reality as spatial, temporal, and causal, which Kant refers to as forms of intuition. We can never have experiences disconnected from these forms of intuition.

There is an inherited preunderstanding in our way of perceiving the world. We interpret the world based on our conceptual frameworks and can never gain knowledge of how things really are. The world we perceive is sensory experiences that we organize into an order that can be understood by using concepts and names for them. There are no objects with necessary properties. What is necessary for something to be as it is, is determined by how we choose to describe it, not by any property in the object. How reality is "in itself" is beyond our concepts.

The closest we can get to a fundamental understanding of reality is by understanding how our structures for understanding work. We can study and clarify concepts, meanings, and representations, but not the world as it is in itself. What we are aware of in our external world are things that are located in time and space. These are things as they appear to us through the lens of our understanding, and they exist independently of us.

Since objects do not exist independently of our senses, we cannot describe the world as it actually is. Space is not a sense-independent world that exists independently of our knowledge of it. Representations are structures imposed on us by our minds. That is all we can have knowledge of. What could lie beyond our instruments for knowledge, we cannot know anything about. Time and space are mental constructions and do not exist in reality.

Everything we perceive and have knowledge of exists in our consciousness. The entire objective world is and remains conditioned by the subject. There is an objective reality that is independent of us and our capacity to perceive, but it is only the human interpretation of reality that we can perceive.

Knowledge is a human creation, and therefore we lack the ability to grasp things in their true conditions. Our knowledge depends on the ability to perceive and interpret the world around us. Our mind structures and organizes experience according to certain fundamental categories. These are innate categories that the mind uses to understand the world we exist in. We can never know how or why our perceptions are limited by these categories, and therefore, we can never fully grasp the "things-in-themselves." Things are independent of ourselves, and we can only know how they appear to our senses. The knowledge we have about things and phenomena is conditioned by ourselves.

Kant makes a distinction between knowledge “a priori” and “a posteriori.” A priori knowledge is knowledge that we have before experience. It comes before experiments and empirical evidence and does not require arguments or proof. It is knowledge that is not based on the world around us but on how our consciousness is organized and functions. It is given by our cognitive faculties and is not a property of things in themselves. An example of this is mathematical knowledge. Since we don´t need any experiences to learn that 2 + 2 = 4, this knowledge is independent of experience. In the mathematics, we use reason and logic to determine what is true. A posteriori knowledge is knowledge that we have after experience.

This knowledge comes after experiments and empirical evidence and requires arguments and proof. It is given by the combination of our cognitive faculties and sensory experiences. For example, we can say that we know something we have learned through our senses, which we have read or heard about.

Kant argues that we must distinguish between the world as it is in itself and the world as it appears to us in our forms of knowledge. Kant does not think of reality as it is but as we perceive it. He calls reality as it is in itself "the thing-in-itself," and reality as we perceive it "the thing-for-us." We know nothing more about the thing-in-itself than its existence and the impressions it gives us, which are the "raw material" for how we organize it according to the principles of time, space, causality, unity, and multiplicity. The thing-in-itself is independent of our interpretation, and the thing-for-us is the human interpretation, filtering, and experience of a thing.

Kant is known for his groundbreaking work in ethics. His ethics is summarized as reason being the foundation of ethics. Human nature and inclinations are irrelevant from an ethical perspective. The good will is the only thing that has absolute value, and duty is of utmost importance. His ethics is based on our rationality as the ultimate good He is convinced that humans are fundamentally rational. He formulates, in his ethics, the categorical imperative, that is the highest principle of morality. It is a moral command that says every person should act in a way that they also wish others to act.

We should not treat another person merely as means to an end, and never treat others for our own purposes, but only with regard to their own well-being. According to this principle, the moral value of any action can be assessed. In his ethics, there are strict requirements for what can be called a "good action." The good action is performed out of duty and is based on reason. Helping someone we care about does not necessarily have to be a good action if it misses our duty to uphold the truth.

After Kant, different areas of research emerged for the objective and the subjective world. The objective world is assigned to the natural science, which focuses on the empirical and the measurable. The subjective world is assigned to the humanities, which focus on human experience of meaning. Kant, as a theorist of knowledge and metaphysics, has had several followers, one of whom was Arthur Schopenhauer.

Arthur Schopenhauer

Arthur Schopenhauer (1788-1860) was a German writer, philosopher, and follower of Kant. Schopenhauer thought that the thing-in-itself couldn´t be cause of phenomena, because causal relationships can only exist in the world of phenomena. Instead, it is about two sides of reality that condition each other. As a phenomenon, the world is a representation, but as a thing-in-itself, it is will.

His book "*The World as Will and Representation*" (2000) begins with the words "the world is my representation." This should be understood as the world we experience being an object for us as subjects, and therefore only existing dependent on a subject. Like Kant, he claims that our knowledge is limited to the world of phenomena. The world of phenomena is a world of temporal, spatial, and causal connections, where all influence each other mutually.

Of all the objects in the world of phenomena, it is merely ourselves we know from the inside. By understanding what ultimately drives us, we can also get a hint of the force behind, not only human actions, but behind all striving and development in the universe. According to Schopenhauer, the origin of the world of phenomena is an absolute and unchanging unity, which he describes as will. It is an impersonal, unconditional, blind will, not a will with motives or goals. For Schopenhauer, phenomena are the same as will in objectified form.

When the will objectifies itself, it does so in accordance with certain determined forms, which can be arranged hierarchically. He speaks of different degrees of the will's objectification, from the lowest, which are the general natural forces, to inorganic matter, plants, animals, and the highest in humans. Since humans have developed a capacity for reflection and self-awareness, the most objectified form of will, is represented by humans.

In his book he conducts an analytical investigation into what is called 'the law of sufficient reason,' according to which everything that happens has a cause. The most essential idea is the distinction between different types of reasons, whether they are called causes, motives, or something else, and the confusion that arises when they are mixed. According to Schopenhauer, nothing exists without a reason. Everything that is has a cause that explains why it exists, and nothing stands without cause on its own.

This principle of sufficient reason can be divided into four different types of causes and thus explanations. The first is causal relationships in our everyday experience and causal explanations made by physics. The second is the explanations we make by referring to something in our experience to prove that a statement is true. The third is the mathematical relationships, we make by deducing a mathematical theorem from another. The fourth is when we seek to understand the reasons or causes behind someone's actions, and explain them in terms of motives. We resort to these four types of causal explanations to understand the world and the actions of others. When we seek to understand what knowledge means and make the world comprehensible, they are useful in different contexts.

When Schopenhauer discusses motives for acts of will, he delves deeper into the will as a central concept in his philosophy. It is not a matter of motives or goals, nor of any relationship between subject and object, or many different wills. It is about a single will that permeates everything and is the core of life and reality. This will has different degrees of consciousness, that can extend from inorganic matter to living organisms, and to the highest stage in humans. Our only way to experience this will is through our own subjective will. The will we experience is not encompassed by the forms of knowledge, time, space, and causality.

Behind the world of phenomena lies the will, blind and constantly striving, with no other goals than its own struggle. It permeates our existence and the whole world, both the inorganic and the organic. The will to live in nature and in us humans is the same will, a world-will.

The world, as a thing-in-itself, is a great will that does not know what it wants, because it only wants and is will. It is beyond all that reason and rationality can understand, but we experience it directly as the most intimate of all.

We can recognize it in our own being, in what blindly and unchangeably drives us towards an unknown goal. We can observe everything else, but our will is something we are intimately familiar with in a completely different way. This will manifests itself as different types of longing and forces in nature, such as gravity, magnetism, and sexuality. It encompasses all seeking, our desires, our hopes, love, fears, hate, contempt, and sorrow. All bodily functions, appetite, digestion, our body as will, are a manifestation of the will to live.

After Kant and Schopenhauer, the central question became whether knowledge was conditioned by an objective reality or by the human subject. The question was whether an objective world exists independently of human perception or not.

Two theories of knowledge with opposing views on this issue are objectivism and subjectivism. Objectivism is a term that refers to the belief that there is an objective world independent of human perception. Knowledge is exclusively conditioned by an objective reality. It is objective, universal, unequivocal, and independent of context. It is an objective reality free from the interference of the content of consciousness. For something to exist, no subject is required. Ethical statements and values are objective and determined by the nature of reality, not by personal attitudes, opinions, or conventions.

Subjectivism is a term that refers to the belief that there is no objective world independent of human perception. Knowledge is conditioned by the human subject, not by an objective reality. There is no objective reality free from the interference of the content of consciousness. For something to exist, a subject is required. Knowledge is subjective, individual, ambiguous, and dependent on context. Ethical statements and values are subjective and determined by personal attitudes, opinions, and conventions. Through self-observation and reflection on our thoughts, we can gain knowledge, insight, and understanding.

An extreme variant of subjectivism is solipsism. Solipsism (from the Latin "solus ipse," alone self) is a term that in an ontological sense denotes the belief that the only thing that exists is one's own experience without underlying causes. In an epistemological sense, it denotes the position that we can only be certain that our own experiences exist. All other beings, all matter in the world, would therefore only be part of one's own life experience.

26. ONTOLOGY AND METAPHYSICS

The question of the fundamental nature of reality is addressed in the fields of ontology and metaphysics. Ontology is the study of being. By starting from being, ontology seeks to understand the fundamental nature of reality. The ontology deals with how the world, or things, fundamentally exist, what their essential characteristics are. An important part of ontology is to describe the properties inherent in a thing's nature (essence) and those that are contingent (not necessary for its nature). If a thing lacks any of the properties inherent in its nature, it does not cease to be what it is. It is merely deficient in some aspect. The thing has an "ontological absence," which means it lacks a part of its being. For example, a blind person has an ontological absence in that sight is part of human nature, but they do not cease to be human because of it.

An important task for the ontology is to determine how different types of things exist and to understand what existence means for different types of entities. Ontology assumes that entities can be divided into different categories. The four categories proposed by ontology include physical, psychological, and abstract entities, as well as phenomena at the limits of perception. Each of these categories exists in its own way. Physical entities are attributed existence based on their persistence in sensory perception. They are tied to specific locations in spacetime. If one thing is in a certain place in spacetime, another thing cannot be in the same place. They can be moved or destroy each other if they try to occupy the same space. The existence of physical entities can be confirmed by perceiving them with different sensory organs or through consistent perception by different witnesses at different times. Some are not directly accessible to our sensory organs but only indirectly accessible through reasoning and inference. In such cases, their existence becomes a mental hypothesis.

Psychological entities, such as thoughts and feelings, have a different kind of existence than physical entities. They have no localization in spacetime, but they exist in the mind. Their existence does not depend on the persistence of sensory perception or different sensory organs. Abstract entities, such as ideas, numbers, and geometric shapes, also have a different kind of existence than physical entities. They exist independently of our perception and thoughts about them. There may be numbers that no human has ever thought of, but their existence is not diminished by that. Their existence is based on possibility within the realm of natural numbers.

Phenomena at the limits of perception involve possible existences. This is where we can discover phenomena that we were not aware of before. Human perception is limited. For example, we can only perceive a small part of the entire spectrum of light, and we can only perceive sounds that have a certain wavelength, neither too short nor too long. But there are animals that can hear and communicate with these sounds because their hearing is different. Their sounds do not exist only as waveforms but also as experiences in their mental world. Our sensory organs are limited, and reality is more than our perception. There are phenomena beyond human perception.

Reality consists of everything we know, everything we do not yet know, and everything we can never know. We can create useful theories and models about the world, but they do not provide any absolute truth about the true nature of the world. A theory is at best a useful map to help us navigate reality. Words, concepts, and theories can only point to a reality beyond words. They reduce reality to a finite representation. Not all knowledge can be analyzed, thought out, and expressed in words.

Metaphysics

Metaphysics refers to something beyond physical reality. Metaphysics deals with things that we cannot perceive with our senses. From originally been dealing only with existential questions, metaphysics now also includes questions about the nature of consciousness, free will, and concepts such as space, time, and causality. Metaphysical statements are not considered part of natural science because their validity cannot be proven or disproven through logical argumentation. Metaphysics doesn´t have to be unscientific. Metaphysics can lead science to new horizons and discoveries. Science is not only about, through scientific method, seeking evidence and counter-evidence, but also about creativity and imagination. Science rests on a metaphysical foundation when it uses creativity and imagination to create new hypotheses.

27. KNOWLEDGE DEVELOPMENT

Scientific knowledge development can be divided into two major areas: natural Science and human Science. Natural Science include subjects such as physics, chemistry, biology, earth sciences, astronomy, and medicine. Human Science include subjects such as humanities, behavioral sciences, and social sciences. Most research fields can be categorized into these major areas, although it can sometimes be difficult to determine the boundary between them. Therefore, they should be viewed more as guidelines than strict boundaries.

A debated issue is whether human Science should use the methodology of natural Science or if they need their own special methodology. The question is whether the study of humans requires a different explanatory model than the study of natural phenomena. Some researchers argue that the methodology of natural Science is the best one and prescribe a methodology for human Science based on natural Science. However, most agree that there is a difference between natural Science and human science. It is the study of humans, human products, and human cultures that constitutes the crucial difference. Natural Science and human Science have different goals and objects of study. Natural Science aim to explain, while human Science aim to understand. The object of study in natural Science is nature, while in human Science, it is humans.

Natural Science

Natural sciences study the material world and the properties and fundamental components of matter. Essentially, it is about particles and their interaction in visible matter. The scope of natural sciences ranges from the subatomic world to the cosmological world. Research objects in natural sciences are special because relevant objects are given quantities in nature and discovered empirically. They have objectively given properties and are not dependent on thoughts, feelings, and actions.

The research methodology applied in natural sciences differs from the methodology applied in human sciences. Natural sciences research methodology is quantitative and requires that results can be generalized. Results should be the same if the investigation is repeated. Truth is unequivocal for natural sciences. A statement can only have one truth value. It can be either true or false and requires reasons and arguments. There must be some evidence or counter-evidence for a statement to be meaningful. Natural sciences use mathematics as a tool to discover and explain regularities in nature and the cosmos. A statement is only meaningful if it can be formulated mathematically and quantitatively. Natural sciences relies on formal logic, which is a very limited and strictly rule-based form of logical reasoning.

Human Science

Human sciences study the meaningful aspects of human experiences and the meaningful structures humans create. There are no meaningful contexts that do not fundamentally involve conscious beings who perceive the context as meaningful in some specific way. Research objects of human sciences are special because object relevance are determined by the meaning that people choose to assign to them. They do not already exist as given quantities that can be discovered empirically. They do not have objectively given properties and depend on our thoughts, feelings, and actions. Studies in human sciences are often normative. The goal is not only to describe what is meaningful to humans but also to change behaviors, cultures, and societies.

The research methodology applied in human Science differs from that applied in natural Science. Methodology of human sciences is primarily qualitative and does not require the results to be generalizable. The results do not need to be the same if the investigation is repeated. For human sciences, truth is ambiguous. A statement can have more than one truth value. It can be both true and false. It does not require reasons or arguments, and it does not need evidence or counter-evidence for a statement to be meaningful.

Human sciences use language as a tool to interpret and to understand the variation in human experience of meaning. All statements that can be formulated in words are meaningful. Formal logic is insufficient to understand the multicausality that humans give rise to. Multicausality is a term that refers to explanations through many factors that are impossible to separate from each other.

Hermeneutics and dialectics are two specific methods commonly used to interpret and understand phenomena meaningful to humans. The word "hermeneutics" comes from the Greek word "hermeneuein," which means to interpret, explain, or announce. Hermeneutics is a research methodology where interpretation is central to understanding human meaning. The word "dialectics" comes from the Greek word "dialegein", which means to separate or the art of conversation. Dialectics is a research methodology where dialogue is central to understanding human meaning. The dialogue can take the form of one partner making a statement or asking a question, which is then met with a yes or no by another partner. Each statement is met with a counterargument. These are combined into a logical synthesis where new insights are formulated. The starting point is that each thought process gives rise to and carries its opposite. It is a method of thinking with opposites. The goal is to reach new insights through dialectical thinking, not to win an argument.

Most agree that there is an apparent difference between natural sciences and human sciences when it applies to the production of knowledge. Natural scientific knowledge is not context-dependent and can be applied generally. Human scientific knowledge is more context-dependent and cannot be applied generally. Unlike human sciences, natural sciences are deductive sciences that deal with prediction through deductive reasoning and mathematical theories. The production of knowledge occurs through empirical experiments, observation, measurement, and data collection to test hypotheses, theories, and laws.

Natural scientists criticize human scientists for having overly broad problem formulations, depend on context that are impossible to draw general conclusions, and a T relativization of the concept of truth, where individuals can claim their own knowledge. Human scientists criticize natural scientists for getting stuck in too narrow problem formulations, disregarding the complexity of subjects related to humans and societal issues. Precision in natural sciences comes at the expense of meaning, leading to only experts being able to claim knowledge.

Several attempts have been made to bridge the gap between natural sciences and human sciences. Various forms of interdisciplinary collaborative projects have been initiated. For example, researchers from different disciplines can collaborate on a joint research project, contributing their various perspectives. A project might involve lactose intolerance, where biologists contribute from a biological perspective, and psychologists from the perspective of how people perceive and manage their lactose intolerance. Researchers in natural sciences and human sciences can be placed in the same institution to facilitate regular meetings and thereby initiate some joint research projects.

28. EDUCATION

Education has been defined as the process through which an individual acquires knowledge and develops an understanding of the cultural conceptual world through teaching and training. The word "education" is derived from the German word "bildung." The term "bildung" was a buzzword in the German-speaking world during the Romantic era, an intellectual movement that emerged in the late 18th century. The terms education and bildung were synonymous but have since taken on different meanings. While education is a goal-oriented process governed by external rules and controlled by authorities, bildung is a voluntary internal process without a concrete goal.

The phenomenon of education is much older and can historically be traced back to ancient Greece and to the Orient and Mediterranean countries. In China, Babylon, and ancient Egypt, there was a need to educate scribes, which possibly gave rise to the first schools, dating back four or five thousand years. In ancient Greece, there were state institutions to educate citizens from the age of seven until the age of military service, at twenty. The purpose was to create a competent military force. They were educated in music, for patriotic reasons, and in physical exercises to achieve physical fitness. Besides patriotic purposes, education also played an ethical and religious role by transmitting prevailing norms.

The Greek philosopher Socrates (469 BC - 399 BC) is regarded as one of the founders of Western philosophy. Socrates is known through the Greek philosopher and mathematician Plato (428 BC - 348 BC). It is in Plato's famous texts "Dialogues" that we learn about Socrates' contributions to education and his theory of knowledge. According to Socrates, education should aim to provide knowledge that leads to insight and awareness. Therefore, a teacher should not impose ready-made ideas on the students or thoughts that are foreign to them. Instead, the teacher should ask questions that can lead the students to the right answers. The theory is based on the ancient belief that nature provides humans with true knowledge, but it is knowledge we have forgotten.

In ancient Greece, there was a need for patriots and soldiers to defend the city-state, and education was organized accordingly. The Romans referred to the subjects taught as the seven liberal arts, which were hierarchically classified based on their value. Plato advocated for an education that serves the state and society by utilizing and developing the inherent qualities that an individual possesses by nature. Based on this, Plato developed an educational system with different types of education for different social classes. Plato's educational system was detailed concerning subjects to be taught, why they should be taught, and consequences education could have for the students and the state. Plato saw education as a deliberate tactic to create the competence the was needed for the state, such as military personnel, merchants, and leaders. With this societal pedagogical perspective, Plato could argue that the state should finance all education,.

Plato founded Plato's Academy, which, together with Aristotle's Lyceum, influenced the science and philosophy of the time. Plato views the education from a societal perspective and can be seen as an early advocate of social pedagogy, while Socrates views education from an individual perspective and can be seen as an early advocate of individual pedagogy. As a guideline in most educational systems, it is established that the degree of individuality should not negatively impact an individual's ability to function in society, and the purpose of education is to adapt the individual to society.

Skills and mental knowledge were taught in similar ways during ancient and classical times. The students could observe how older generations did things, imitated them, or learned through memorization. This teaching method is often referred to as imitation or mechanical learning. Imitation still has a place in education, such as learning pronunciation in foreign languages or learning techniques demonstrated through others' experiences. Practice and repetition were considered to enhance memory retention and therefore deemed essential in education. However, imitation leaves no room for individuality, independent thinking, and creativity.

Lecture-based learning is another ancient method similar to the aforementioned teaching methods. In lectures, the teacher reads or narrates while the students take notes. The emergence of the lecture method can be explained by the scarcity of books in ancient times. When books became more common, reading aloud was replaced by commentary and explanations of textbooks. Lectures provided opportunities to ask questions about textbooks. Since then, abstract learning, whenever possible, has been replaced with visualization and concretization, partly motivated by the fact that more people have access to education, including those who struggle with abstractions. Various methods have emerged, focusing on the learner, and direct instruction has decreased.

In traditional education, it was important to find answers to factual questions, but there were no requirements to understand and explain. Commonly, it only required searching for facts and the ability to describe them. The questions in textbooks usually dealt with what, where, when, and how many. There were rarely questions that encouraged seeking explanations. It was the teacher's responsibility to search for and provide explanations. The traditional view of pedagogy was teacher-centered, with the teacher presenting knowledge and explaining concepts. Before the 1970s, behaviorism was foundation of education. Subsequently, there was an emphasis on individual thinking. Factual questions and memorization of facts were no longer primary focus and understanding and explanation became the new focus. Interest shifted towards understanding cognitive abilities and mental thought processes, and form the basis for cognitive teaching methods. Teaching methods can be described in three major areas called behaviorism, cognitivism, and constructivism. All have had a significant influence on how teaching should be conducted in our schools.

Behaviorism

Behaviorism is a psychological approach that has focus on behavior and learning, with its roots in the early 20th century philosophical and scientific currents. Behaviorism emerged as a reaction to theories that used introspection as a method to describe human behavior. According to behaviorism, only observable behavior could provide relevant measurements, and only physiological and biological behaviors should be studied. Behaviorism reached its peak during the first half of the 20th century. Most learning research during this period focused on observing and objectively measuring responses that individuals exhibited to external stimuli. The inner workings of the brain were not measurable and were therefore considered unimportant.

A fundamental idea in behaviorism is that consequences affect behavior if they are repeated and depending on the intensity with which this happens. Consequences can be pleasant and desirable or unpleasant and undesirable. This reinforces or weakens the likelihood that the same behavior will occur in similar situations. A positive reinforcement involves a positive stimulus that increases the strength of a certain behavior. Negative reinforcement involves a negative stimulus but has the same goal as positive reinforcement, which is to increase the strength of a behavior. In a negative reinforcement, unwanted behavior can be reduced through punishment, extinction, or "time-out". Punishment means that an unpleasant, unwanted stimulus is given immediately after unwanted behavior. Extinction means ignoring the behavior. Time-out means temporarily removing a positive reinforcement.

In an educational context, positive reinforcement can be receiving attention and praise from the teacher, good grades, or other types of rewards. To make effective praise , it should be specifically linked to the behavior being reinforced. A teacher can reinforce a certain behavior consciously or unconsciously. The idea is that motivation increases through the reinforcement of desired behavior. The individual is rewarded for their achievements, which raises motivation. Individuals learn new behavior faster if they receive reinforcement for each correct response. It is the teacher who controls and defines what the learner should achieve. The premise is that knowledge is something that is transferred to the individual from external sources.

Behaviorism has its roots in classical and operant conditioning. Classical conditioning involves reflexive and automatic behavior. The behavior occurs regardless of consequences because it is an involuntary response. Operant conditioning is a learned behavior that can mentally control the consequences of the behavior. There can be more of a behavior that is reinforced and less of a behavior that is punished. As an example, experiments have shown that desired behavior in rats is reinforced when they receive a reward for doing things correctly.

Cognitivism

In the 1950s, the cognitive revolution began, leading to the questioning of behaviorism. Unlike behaviorism's research interest, the cognitive tradition focused on mental processes such as thinking and intellect. Studies focused on the brain's cognitive processes, such as perception, memory, problem-solving, creativity, and information processing. Humans were not longer seen as passive entities that could be conditioned to learn, as in behaviorism. Instead, individuals were considered driven by conscious intentions and motives. Cognitive maturity was significant for the ability to assimilate different types of knowledge.

The basic idea in cognitivism is that individuals have the

drive to seek knowledge and experiences. Individuals are proactive in that they act based on their experiences rather than being products of experiences. The individual is guided by awareness of purposes and motives, not by emotions. It is the individual's desire to understand their environment that acts as motivation. Cognitivism focuses on describing and understanding cognitive abilities and mental thought processes. The focus is on the cognitive ability to acquire and develop knowledge. The ability to analyze and think rationally is considered higher-level skills. Knowledge is part of cognition as a whole, is something the individual must construct themselves, and linked to understanding. Understanding is the result of the individual's own thought activity, when the individual integrates and forms an opinion about something.

Cognitivism is based on three assumptions of learning.. Firstly learning is about the construction of knowledge rather than the registration of knowledge. Secondly, learning depends on prior knowledge, meaning that existing knowledge is used to create new knowledge. Thirdly, learning depends on the situation in which learning occurs. Knowledge is not something that is transferred to the individual from external sources, and the teacher does not control and define what the learner should achieve. What the learner already knows determines what should be noticed, remembered, and learned. The most important element in learning is what the learner brings to their learning situation. Therefore, the environment should create conditions for learning but without being controlling. The focus is on the best way to support the individual's learning ability.

Constructivism

Constructivism emphasizes the active construction of understanding.Two main forms of constructivism are the so called individualistic constructivism and sociocultural constructivism. Individualistic constructivism focuses on how the individual constructs knowledge based on the creation of internal knowledge structures through the individual's experience and self-regulation. It emphasizes how the individual can create understanding through experience and how our brain processes information. The goal of learning is the development of one's knowledge. The premise is that knowledge exists within and depends on the individual and that there are universal forms of knowledge structures that govern our understanding.

Sociocultural constructivism focuses on the construction of knowledge based on social interaction, creating shared understanding through shared experiences. The social and cultural circumstances surrounding an individual are emphasized. The goal is to create opportunities for the individual to develope knowledge. The premise is that knowledge cannot be created independently of the social context in which the individual is located, and that there are no universal forms of knowledge.

Knowledge is a construction that follows from the individual's interaction with the social environment. The social and cultural circumstances around an individual shape the opportunity for development. People are born and develop within the framework of interaction with other people. It rests on the assumption that knowledge cannot be transferred from a teacher to the learners. When knowledge is transferred from one person to another person, something of the knowledge is always lost in the transfer. Knowledge is the result of personal interpretation of experiences, influenced by factors such as age, gender, and existing knowledge base.

According to constructivism, meaningful learning is the active creation of knowledge structures that are based on personal experiences. Meaningful learning occurs when the learner selects relevant information and interprets it using existing knowledge. It is a process that involves selection, organization, and integration. The individual organizes the information into a coherent whole, and integrates it with existing knowledge, and constructs their understanding.

Evaluation is based on how the learner structures and processes knowledge rather than measuring how much the learner has learned. Learning requires self-regulation, and the learner builds concepts through reflection and abstraction. We do not solve problems by memorizing the correct answers. To solve a problem intelligently, we must first see the problem as our own problem. We need to see the problem as an obstacle blocking our path to a goal and have a desire to reach a specific goal. Seeking a solution and finding a way to reach the final goal provides greater satisfaction than the goal itself.

The educator is merely a discussion leader in the process of constructing understanding. The discussion leader's role is to support and challenge the learners to build knowledge and understand context. The role is to help the learner develop skills in learning and use these skills to construct more useful knowledge. The learner should be helped to develop thinking and learning strategies that are relevant. By building on what the learner already knows and providing appropriately challenging tasks, a teacher can support the learning process at different stages of development. The teacher should guide the learners' activities so that they discover and construct a personal and meaningful understanding of the problem or subject.

To support constructivist learning, the teaching situation should provide opportunities to be exposed to different perspectives on a problem. One way to support the construction of understanding is by allowing the learners to see ideas and problems from multiple perspectives. Meaningful discussions and systematic discussions are helpful for creating a personal view of the subject being learned, and discussions support the construction of knowledge. The teacher can contribute to discussions among the learners by providing a provocative idea related to the subject. Education shifted to a more process-oriented pedagogy.

Problem-Based Learning (PBL) is an example of process-oriented pedagogy based on constructivism. A central principle in PBL is that learning should be based on real-life situations and a realistic context. A realistic context is an environment where learners solve realistic problems using their skills and knowledge. Another central principle in PBL is that the learner should take responsibility for their learning and drive their own learning with the support of peers and teachers. Often, this means that the learner searches for facts and can describe them but lacks the ability for critical thinking to make judgments about what is relevant and meaningful. The teacher's role is to promote discussions and the independent thinking among the learners rather than being the one who holds the knowledge. The task is to support the learner in building knowledge and context understanding. It is a way to stage education based on the philosophical and theoretical assumption that the learners own activity before and during group work is crucial for learning.

Generative AI

Generative AI is a branch of artificial intelligence that has transformed our perspective on education and society, particularly since the AI-based language model ChatGPT was introduced as a prototype to the public in 2022. It is a virtual assistant that can answer questions of fact, and generate useful texts of various genres in many various languages. ChatGPT gained attention for its detailed and articulate responses across different fields of knowledge. It is designed to interact with users, answer questions, provide information, and engage in conversations on various topics.

Within the context of education, generative AI has been called a revolution and is said to be the most significant development since internet development. Generative AI is a collection of algorithms and techniques that train computers to create content, whether it be text, images, music, or other things that humans typically create. By training on large amounts of data, these AI models can learn to generate content similar to what they have been trained on. However, their knowledge is limited to what they have learned from their training data.

Generative AI has the potential to transform education and the educational system in many interesting ways. With the help of this technology, a more personalized and efficient learning environment can be created for all students, regardless of background or abilities, by giving each student a chance to reach full potential. Various examples are proposed how generative AI can improve the quality of education and be applied in the educational system.

In the context of learning, it can create individualized exercises, providing different teaching materials that help the students improve their skills and start from their own level. Generative AI can stimulate the student's creativity and skills of solving problem by generating various scenarios and challenges. AI-driven teaching assistants can alleviate teacher's workloads by answering common student's questions, planning lessons, and monitoring student's progress. The assistants can provide immediate feedback on student's work, which can help them to improve their skills. They can offer students realistic environments to explore and learn through AI-driven simulations and virtual worlds.

It is important to understand that these language models do not always have the correct information in the content they produce, as this is not their primary purpose. The students should have the possibility to explore and test how these types of language models work so that they understand how they are used. They need to gain insight into both advantages and limitations of the technology. The development and uses of these language models lead to the question about the value of knowledge.

29. THE VALUE OF KNOWLEDGE

In the Western knowledge tradition, it was taken for granted until the 20th century that knowledge was something only humans could produce. With the development of artificial intelligence and Freud's theory of the unconscious, the discussion of human and non-human knowledge arose. It was no longer unusual for knowledge to exist in biological systems, AI systems, immune systems, and DNA. Knowledge could be generated by any biological or technical system that made a difference in reaction and behavior. Knowledge also encompassed animals' instinctive behaviors and plants' reactions to light and darkness.

Both AI researchers and biologists claimed that knowledge was no longer something only humans produced, but that AI systems and biological systems could also produce it. Intelligent machines and biological organisms can learn to generate new knowledge without consciousness. They can learn to interact with the external world, store, and process information without consciousness. To create a definition that covers this type of knowledge, biologists have proposed three criteria: the AI system must be self-organizing and dynamic, the knowledge must represent the external world, and the system must have access to the information quickly enough to be useful.

The Israeli auther and professor of history Yuval Harari writes in his book "*Homo Deus: A Brief History of Tomorrow*" (2017) that the value of knowledge lies in its ability to lead to an improvement in human behavior and society in some respect. Knowledge that does not lead to any change is just history - a reminiscence of the past. It becomes a container for what we already know. There is a paradox with historical knowledge because it quickly loses its relevance. The more data we have, the faster our knowledge becomes outdated. In our attempt to understand what is happening, the accumulation of knowledge accelerates, which has the consequence that we become increasingly unable to make predictions for the future and to make meaningful use of the present.

Traditional paradigms for the development of knowledge and education are no longer sustainable. The primary focus is no longer on acquiring as much information and facts as possible but on being able to sift through all the information and determine what is meaningful. Changes occur so rapidly that it is impossible to know what skills and knowledge will be required in the future.

The development of scientific knowledge increases exponentially and tends to move towards greater fragmentation and specialization. Each research area develops its own methods and specialized language, making it difficult to get an overview and integration of knowledge from different fields, which is required for groundbreaking creative breakthroughs.

Science and education, focused on first acquiring all the knowledge within a specific field, are no longer possible. The growth of knowledge is so large and changes so quickly that it is impossible for researchers to follow all knowledge development within their own field, let alone follow it in other fields. It is impossible to monitor the questions at the forefront of research within one's own area and the questions at the forefront of research in other areas. This can explain why groundbreaking creative breakthroughs are rare.

30. SELF-KNOWLEDGE

All development of human knowledge has its origin in the human mind. Everything that our mind can know and be in contact with is its content, and therefore it is important to understand the nature of our own mind. Methods used for self-knowledge in Western and Eastern cultures are different, but both are focused on the development of self-understanding. The difference can be significant, which may be due to historical, philosophical, and religious factors.

Methods for self-knowledge in Western cultures have their roots in philosophical traditions such as rationalism and empiricism, where self-analysis plays a central role. Eastern culture, methods for self-knowledge have their roots in philosophical traditions such as Hinduism, Buddhism, and Taoism, where meditation plays a central role, especially in countries like China, Japan, and India.

One difference is that Eastern cultures place greater emphasis on achieving inner peace and harmony, while Western cultures are more focused on analyzing the personality. In Western tradition, the "self" is an individual entity separated from the external world, while in Eastern tradition, the "self" is not separated from the external world but is part of a larger cosmic order. Western methods for self-knowledge are aimed at achieving specific goals in a limited time, while Eastern methods are focused on long-term development over time.

In Western cultures, psychotherapy is a method to help people deal with and overcome mental problems, emotional and behavioral difficulties. It is a form of treatment for therapeutic interaction, where a therapist works with an individual or a group to explore thoughts, feelings, behaviors, and identify problems.

The goal of psychotherapy is to reduce symptoms of mental illness, improve the individual's relationships and function in everyday life, promote personal development, and increase quality of life. Psychotherapy can be short-term or long-term depending on the needs and severity of the individual's problems. There are many different forms of psychotherapy, such as psychodynamic therapy, cognitive therapy, gestalt therapy, family therapy, and many others. Each therapeutic form has its theoretical assumptions, methods, and approaches, but they often share a fundamental aim to promote understanding, healing, and growth in the individual. Therapy can be conducted individually or in groups, depending on the situation and the client's preferences. Psychotherapy methods vary but often include dialogue, self-reflection, analysis of thoughts, emotions, and behaviors. Two common forms of psychotherapy are psychodynamic therapy and cognitive therapy.

Psychodynamic therapy

Psychodynamic therapy is a collective term for forms of therapy based on psychoanalytic theory. The basic idea is that we have an unconscious inner life shaped in connection with other close and important relationships. Mental problems are rooted in unconscious defense mechanisms used against disturbing inner conflicts and emotions. Cooperation with the psychotherapist is a basic prerequisite in meeting the individual. It is a joint exploration of how and why the individual has ended up in their situation. The goal is to understand individuals' problems without judging and to provide the individual with insights into their functioning, fears, and defenses, and possibility to find solutions. Some schools emphasize the interpretation of the individual's transference of feelings and attitudes from past relationships, while other schools emphasize supportive and affirming therapeutic relationship.

The starting point is that there is unconscious mental material that is not directly accessible for introspection and reflection. Unconscious defense mechanisms prevent conflict-laden material from reaching a conscious level. Mental problems originate in conflicts between the conscious and the unconscious. By becoming aware of repressed needs and motives, the individual can change their behavior to become more purposeful.

The unconscious represents hidden aspects of ourselves that influence our conscious and daily lives. Repressed painful memories can coalesce to form specific complexes of memories that live their own lives. This means that the pain we experience can merge with pain from the past and can remain in the body. These complexes are trapped life energy, temporarily separated from our overall energy field. They have autonomous power and the ability to unconsciously influence our thinking and behavior without our noticing. An active complex can temporarily exert a compelling force, leading to compulsive thoughts and actions. These unconscious tendencies can be stronger than our conscious self and can act against our will. They can function as "sub-personalities," which sometimes act as if they have their own will.

A complex can be dormant or active, and anything can awaken a complex, especially situations that remind us of a previous pain pattern. We may have invested our whole self, or parts of ourselves, in a pain pattern that survives through our unconscious identification with it. This means that we have created an unhappy self and believe that this mental construct is our self. However, it is our thoughts and behaviors that keep the pain alive. There is significant resistance to change within us, based on an unconscious fear of losing our identity. We prefer to experience the pain rather than take a step into the unknown and risk losing an unhappy self.

We can observe the resistance and the strange pleasure that can exist in being unhappy. Once it has taken over us, we want more pain and inflict pain on others. It makes us either victims or perpetrators. By adopting a victim identity, we attract circumstances and interpret situations that confirm our victim identity. Instead, we can acknowledge and accept painful feelings as they are. We can try to see situations without resisting what is happening and accept life as it is, even if it is not as we think or wish it were. Painful emotions and thoughts hinder us from accessing the life energy that exists in the body.

Unwanted emotions and thoughts cannot be suppressed, but they can be made less intrusive. By changing our attitude towards life, we can better manage our intrusive emotions and thoughts. We can accept them because they already exist and do not need to project them onto others. We can be aware of them without analyzing, judging, or controlling them, and without trying to deny, fight against, or cling to them. We can view thoughts and emotions as independent objects and not identify with them. This means letting go of dissatisfaction, blame, self-pity, and bitterness that serve no other purpose than to reinforce our sense of an unhappy self. It must be a genuine acceptance; otherwise, the resistance remains and reinforces our sense of being a victim. It only legitimizes a victim identity and becomes a mental stamp to continue being unhappy.

Acceptance has been found helpful in managing psychological suffering. Psychological suffering refers to suffering that arises from our thoughts about painful experiences. Negative thoughts before and during pain increase the experience of pain. When we identify with our pain-related thoughts, psychological suffering arises. By not identifying with thoughts about pain, experience of pain decreases. This means distancing ourselves from pain-related thoughts.

We can manage our pain by experiencing what is happening moment by moment without thinking. By observing the pain in the body moment by moment, thoughts no longer have power over our experience of pain. We can observe the pain, feel its raw energy in the body, and accept it. We do not let the feeling turn into thinking but experience it fully. This reduces the pain and applies to all emotions from anger, fear, and sorrow to a sense of discomfort.

Feelings are not the same as emotions. Feelings are perceptions of emotions, the mental associations and reactions to emotions, and interpretation of an emotional situation. Feelings follow thinking and can play a role in confirming something we have already thought.Feelings arise as a result of memories and are a reflection of thinking. For example, when we think we are threatened, a feeling of tension arises in the body that we call fear, and when we are attacked, a feeling arises that we call anger. Normally, a feeling has a short lifespan, but it can survive for days and weeks.

The word emotion can be traced to the Latin "emovere," which means to disturb. When we have emotions that disturb, we are not in a position to analyze them, but we can feel the energy in the body. Emotions precede feelings, are temporary, unconscious reactions, and the body's response to an event. They are instinctive, visible in body language, such as facial expressions. Emotional reactions are biochemical reactions programmed by our past experiences and the situation we are in. They are a driving force behind our behavior and play a significant role in how we interact and experience life. We are often unaware of how emotions control our lives.

Cognitive Therapy

Cognitive therapy is a form of psychotherapy established in the 1960s. The cognitive theory is based on the assumption that psychological problems and disorders originate from dysfunctional thinking. Together with the therapist, the individual examines and identifies dysfunctional thinking and behavior. The goal is for the individual to understand themselves and become their own therapist. Two variations of cognitive therapy are cognitive behavioral therapy and metacognitive therapy.

The term cognitive-behavioral therapy (CBT) often has the same meaning as cognitive therapy. One difference is that CBT focuses on changing behaviors through learning and training, while cognitive therapy focuses on gaining insight by finding rational solutions to problems. CBT focuses on individual thoughts rather than the processes behind thoughts. It involves mapping separate thoughts and systematically questioning them. By organizing our thinking more rationally, we can change our behavior to become more purposeful. This rests on the assumption that thinking can control behavior and is accessible to introspection and reflection. Thoughts, feelings and behaviors are interconnected and influence each other. Feelings are hard to influence, so the treatment focuses on influencing thoughts or behavior. To achieve new behavior, a relearning must occur through experience and practice.

Metacognitive therapy

Metacognitive therapy focuses on processes behind thoughts rather than on separate thoughts wtth the aim to change our relationship to our thoughts. By our ability to metacognition, we can reflect on our thoughts and intellectually analyze our mental content. We can map thoughts that are unproductive, willfully direct thoughts toward fruitful paths. We can assess the consequences of our actions, and distinguish between functional and non-functional actions relative to a desired goal. By examining the reasonableness of our thoughts, we can dismiss thoughts that do not lead to the desired goal. We can learn to become better at reflecting, planning, and assessing what is reasonable.

Focusing too much on dysfunctional thoughts can result in getting stuck in thoughts about what has gone wrong, could have gone wrong, or might go wrong. This can often be counterproductive by prolonging psychological suffering's reaction to anxious thoughts. It has the effect to increase attention to anxious thoughts, making them more intrusive. It rests on the assumption that anxious thoughts are helpful to manage problems. In the therapy various attention exercises can be used training ability to be focused, attentive, and flexible in what is focused on. These exercises have shown reduced frequency of intrusive thoughts, anxiety, stress, and depression.

Indian Culture

In Indian culture, methods for self-knowledge are written in ancient Vedic scriptures, which can be dated from 1900-1100 BCE. The Upanishads are a special class of philosophical writings in Indian literature that contain the earliest traces of philosophical thinking and form a foundation for Vedanta, which denotes a direction within Hindu philosophy focused on reading and analyzing older religious literature. There are different branches within Vedanta, and each branch has its interpretation of these writings. One of the leading branches in Vedanta is Vedanta Advaita, which means non-duality in Sanskrit.

Yoga is a fundamental concept in Indian philosophy and ancient Indian tradition. Yoga is a term that refers to contemplative and physical techniques. The earliest traces of yoga have been found in stone seals engraved with figures in classic yoga postures, showing that yoga was part of Indian culture as early as 5000 years ago. Historically, yoga can be traced back several thousand years in ancient Vedic scriptures written in Sanskrit. Sanskrit is a poetic language intended to remind us of our origins and how life's transience can be managed through songs and hymns. The word yoga means "to unite" in Sanskrit, referring to the union of body and mind and the union of individual consciousness with universal consciousness.

In the 20th century, yoga began to be made available to a broader public and came to be practiced by people worldwide. In the West, yoga has increased in popularity since the late 20th century. In Western contexts, yoga is mainly associated with physical exercises, the various positions or asanas that constitute part of hatha yoga. Hatha yoga is a term aimed at achieving full control over the mind and body through various physical postures. A common misconception is that yoga is solely a system for physical exercises and postures. In traditional Indian and Vedic context, yoga is more than just physical exercises.

Yoga is a philosophical and spiritual path that includes meditation and yoga exercises, ethical life rules, self-discipline, yoga postures that promote physical flexibility and health, various breathing exercises, and control of life energy. Yoga is a holistic approach to living with greater awareness, balance, and harmony. By practicing yoga, we can become aware of different ways of being. Basically, yoga is about understanding how our thoughts and feelings affect the state of our mind and that we can influence our mental state. Breathing plays an important role in this. Breathing reflects emotional states, and our emotional states are influenced by breathing. Physical and mental health is achieved by balance between the mind and the body's energy system. It is about being in a state of complete mental presence and awareness.

By practicing various yoga exercises, a person can be prepared for the insight into their unity with what is called pure consciousness, and become enlightened. This can only happen when the mind is freed from all distractions and conditioned ways of thinking and perceiving. It is the highest state that can be achieved with a mind bound to the human body. Pure consciousness is invisible, formless, and unlimited. It cannot be understood with the mind because our senses are limited to the visible and forms. We can only experience the world as it appears through our sensory organs. The brain creates the world based on the inflow of data to our senses and cannot provide true information about reality.

Pure consciousness involves the presence of pure being and is the innermost essence of our existence, and the same for all people and other living beings. It permeates everything in nature and manifests individually in human brain and body. Human consciousness is reflections of pure consciousness. It can be likened to how a moon can appear as multiple moons when reflected in different water vessels.

Pure consciousness manifests as duality in our ordinary waking state. When we are in the waking state, we experience a self that is separate from an external world. In pure consciousness, there is no distinction between the self and the external world. When our illusion of a self separated from an external world disappears, there is an experience of our unity with all living things. This changes the perception of ourselvs and the world.

Satchitananda refers to the subjective experience of pure consciousness. Satchitananda is composed of the three words sat, chit, ananda, which are inseparable aspects of pure consciousness. Sat refers to what is enduring and real. It is the container for all our experiences. Chit refers to what we are aware of, our experiences in usual waking consciousness, such as names, forms, and knowledge. Ananda refers to an inherent quality of lasting happiness, that is distinct from happiness we experience of transient pleasures. Ananda also includes principles of movement and change.

31. HAPPINESS

Western culture differs from Eastern culture in its views on what brings happiness. Western culture is more inclined to study the causes of unhappiness than what brings happiness. All people seek happiness, yet there is more research on what makes us unhappy than on what makes us happy.

Stefan Klein

Stefan Klein is a German physicist who has studied philosophy, conducted research in biophysics, and worked as a science editor at the magazine Der Spiegel. Klein has compiled research from one of the world's largest brain imaging archives, established by the Portuguese neurologist Antonio Damasio and his team. Through CT scans, over 2,500 living brains have been examined there. Thanks to this image archive, Klein has shown how joy, disgust, fear, and anger originate in the brain. Klein provides a summary of how neuroscience views happiness in his book "*The Science of Happiness: How Our Brains Make Us Happy and What We Can Do to Get Happier*"(2015), where he links happiness to the brain's reward system.

Reward system is a term for the network in the central nervous system responsible for regulating and controlling behavior by influencing feelings of pleasure. It involves several automatic brain structures, which are versatile and adaptable. We are equipped with several different systems for our desires and there are different systems for pleasant and unpleasant affects. These can work in parallel or counteract each other.

What we associate with happiness depend on culture, upbringing, and media advertising, which tells us what we should do to be happy. When we believe something makes us happy, we want more of it and we create a dependency. Happiness only has reality as long as we believe in it. It is a psychological reward that reinforces behavior and operates on a biochemical level through neurotransmitters.

A central theme in the book is that happiness originates in our body, thoughts, and beliefs. Our body is connected to our well-being and general psychological state and is influenced by unconscious emotional reactions in the body. When our emotions are connected to the right body signals, we can experience well-being. This occurs when the brain receives the right signals from the heart, skin, and muscles. Through these signals, our perception of physical well-being is constructed. Research shows that the feeling of happiness and unhappiness has different brain connections and chemistry.

Two discoveries in neuroscience have been crucial for research on happiness. The first discovery is that we are programmed to experience positive emotions in the same way we are born with the ability to speak. The second is that even an adult's brain can continue to change. This means we can cultivate our positive emotions and that if we consciously engage in something, it becomes more deeply rooted in the brain. Connections for positive emotions can be strengthened through conscious exercises. Every time we learn something, the brain changes. The brain has the ability to be plastic, so we can train ourselves to manage our emotions.

When we engage with activities that generate positive emotional experiences, the brain is shaped in a positive direction. Feelings of happiness are no coincidences but a result of the right thoughts and actions. When we are oriented towards happiness, we seek situations that make us happy. Therefore, we should primarily aim to change ourselves, not our circumstances. There is no doubt that we can learn to be happy, says Klein.

We can consciously control negative emotions. We do not need to act on them in the belief that they will disappear because they do not. Vitality is maintained by appreciating things that cannot be predicted and by trying to see the familiar from new perspectives. An alert mind can increase well-being even when it merely observes, and we can enjoy just being attentive. Happiness lies in controlling negative emotions and feelings.

Controlling feelings is not something that is associated with psychotherapy. It is rather seen as an obstacle to conducting psychotherapy. Some forms of psychotherapy are based on the asumption that expressing negative feelings is helpful for liberation from them. Behind this idea is a notion, dating from the 19th century, that the brain was like a steam engine and that negative emotions would build up excess pressure that must be released to avoid a dangerous overreaction, says Klein.

We carry emotions as a legacy from evolution. Our brains and our emotions developed simultaneously during the evolution. All emotions fundamentally originate from sensory movements, such as desires or fears, which existed in the animal kingdom long before humans. Therefore, it is important to know oneself when seeking happiness. No preparations are needed. It is enough to be attentive to our reactions and habits to discover what makes us feel good.

32. REALITY

Reality (Lat. actualitas, realitas) is a concept that denotes both the quality of being real and everything that is real. Beyond this linguistic distinction, two main meanings can be identified. Firstly, it can denote everything that is actualized, as opposed to merely possible, or the actualization itself as opposed to the mere possibility. Secondly, it can be contrasted with the merely apparent, that is, reality as it falsely appears to an observer.

There is a classic opposition between the asumption that reality is what our sensory impressions give immediate contact with and the asumption that what the senses show us is an illusion or a veil that conceals a true reality.

In the Western world, this thought can be traced primarily to Parmenides and Plato. According to Parmenides, true reality is a homogeneous unity, while for Plato, it consists of a higher timeless, eternal world of conceptual ideas that function as archetypes for the changing reality of the senses.

These ideas are reminiscent of Hindu Advaita Vedanta philosophy with its asumption of the sensory world as a veil of Maya. This line of thought reappeared in different ways later in Kant, who distinguished between the phenomenal world and the world as it really is. Kant also assumed these two are entirely different but added that we cannot know anything about the world in itself.

We perceive reality as divided into two parts: the world inside the body and the world outside the body. The world inside the body is the mental world, containing all thoughts, emotions, and sensory perceptions. The world outside the body is the physical world, with all physical objects. This division is an illusion created by our brain. All experiences of sensory perceptions, feelings, and thoughts happen in our mind. Therefore, exploring the nature of reality should begin with understanding the nature of the mind. We should first understand the nature of our own mind to understand the nature of anything experienced by our mind.

The reality around us consists of electromagnetic waves and light. These are patterns of light transmitted through electromagnetic radiation to receptors in our brain. Energy is projected into the brain to be transformed into something that can be interpreted and understood. This starts with the activation of a receptor, which is then sent for interpretation in the brain. We become aware of the world through the brain's translation of light and waves into something we can know.

It is impossible to explain what ultimate reality is, but we can describe it in categories of physical level, biological level, and level of consciousness. These are qualitatively distinct levels. Each level adds something new that can´t be reduced to a lower level.

The physical level describes the basis of reality as matter and energy. Matter and energy builds up the different systems in the physical universe. Besides matter and energy, there is also information. Information handles instructions and physical mechanisms. Physicists start from matter/energy as given facts when they seek to explain the structure and laws of the universe.

The biological level adds something new. Biology describes the structure and processes of organisms and their life. Life is about reproduction and the transfer of information stored in our DNA. Without information, an organism could not survive and reproduce. Biologi can´t be reduced to the level of physics and information. Biologists start from life as a given fact when explaining the structures and processes of organisms.

The level of consciousness adds something further new and is more complex. Without consciousness, there are no experiences. Consciousness can´t be reduced to something other than itself. Psychologists start from consciousness as a given fact when explaining mental structures and processes.

33. MATHEMATICS AND WORLD OF EXPERIENCE

The perspective on the relationship between the mathematical world and the world of experience has changed over time in the history of mathematics. In ancient Greece, the philosopher Plato (428 BC - 348 BC) claimed that mathematics was superior to the world of experience and the highest form of knowledge about reality.

For the Greeks, mathematics belonged to the Platonic realm of eternal forms which existed in a transcendent world. In the 16th century, the Italian physicist Galileo Galilei (1564-1642) claimed that mathematics was the same language found in nature and the only source of true knowledge. Mathematics provided a secure and well-founded understanding of nature, while the world of experience merely was opinions that were in constant change and impossible to gain knowledge of.

In the 19th century, the empiricism came with a reversed view of mathematics. The mathematics was considered to be a language invented by humans. It was a tool for science rather than absolutely certain and well-founded knowledge. The only thing that provided certain and well-founded knowledge was empirical science. Mathematical and empirical science could coexist but were no longer dependent on each other.

From being the highest form of certain knowledge, mathematics became dissociated from physical reality and fragmented into hundreds of branches, each building on its system of axioms. This had results that it became difficult for mathematicians from different branches, to communicate with each other. At the same time, pure mathematics emerged, operating independently of the physical world. This meant that mathematics shifted from being about concepts related to the experience of the physical world to being about objects without connection to the physical world. Many mathematical concepts arose with little or no physical meaning, and mathematics was introduced as a legitim object for mathematical studies.

In the first half of the 20th century, set theory gained significant importance. It created an independent mathematical world and became a common foundation for all of mathematics. Set theory provided the ability to see connections between different branches of mathematics and made the fundamental assumptions within these branches visible. Concepts that had been vague and unclear could be precisely defined using set theory. It provided all branches with a unified framework and language in which mathematics could be interpreted. Set theory became a standard axiomatic system for various mathematical axiomatic systems. Each system could be expressed in the language of set theory. Mathematics transformed into a science of diversity and pluralism. Today, there are hundreds of formal systems based on mathematics.

Mathematics has led to new technological inventions and has been the foundation for the development of trade and the internet. Our modern world is built on mathematics, all from computers to our understanding of biology and the universe. It has been used in biology to explain the structure and behaviors of organisms and been used in physics to explain the structure and laws of the universe. Mathematics is present in our society and shapes our perception and conceptual understanding of the world. It influences how we think and understand, even beyond empirical investigation, but it does not provide answers to questions of meaning, ethics, and morality.

Mathematics has become an ideal for objective truth and knowledge, even in domains other than physics. It has become a benchmark for all areas of scientific discourse that seek to assert objectivity. By demonstrating their domains to be as mathematical as possible, they are considered highly credible.

34. PHYSICS AND MATHEMATICS

In physics, mathematics is an extraordinary tool for exploring the universe. Mathematics can reveal hidden corners of the universe that we cannot discover solely through our senses. With the help of mathematics, it has been possible to create instruments to uncover the innermost properties of matter. Mathematics provides a tool for precision and rigor that humans alone do not possess. With mathematics, we can determine the properties of particles and the position and movement of stars. Mathematics allows us to discover new patterns and make predictions without the need for empirical data. It can both discover and create new patterns. Once a pattern is discovered, mathematics can further develop and calculate new patterns.

Many discoveries in physics have been preceded by predicting relationships on a mathematical basis, which are then verified empirically. Mathematics is an important tool for formulating hypotheses, testing theories, and summarizing results. In physical research, the theories used to explain the physical world are all mathematical, but they are not identical to the physical world.

Mathematics is an abstract and general science for problem-solving. It is abstract because it is freed from the concrete origins of problems, and it is general because it can be applied to a wide range of areas. Mathematics is a special world that exists independently of physical things in this world. Mathematics is not the same as the physical world because it is not based on empirical evidence, but it can free us from false theories about the physical world.

Physics theories are based on mathematics but have no explanation for what mathematics is or where it comes from. There should be some explanation for what mathematics is and where it comes from, as is the case in other areas of research. Mathematics has never been questioned, despite not meeting these criteria. Physics can mathematically describe the structure of physical reality but makes no claim that mathematics is secure.

A debated question is whether the pure logic that exists in mathematics is secure. For mathematics to be secure, it must be consistent, and statements cannot be both true and false. Mathematics follows strict rules for logical inference that are called proof if it is based on theorems or axioms. There are mathematical statements where it is not possible to determine what is true and false. David Hilbert (1862-1943) and Kurt Gödel (1906-1978) are two logicians with different views on whether the pure logic is secure.

David Hilbert

David Hilbert is a German mathematics professor known

as one of the founders of proof theory and mathematical logic. In 1920, he started a fameous program, aiming for a complete consistent axiomatic system for mathematics. This program became the starting point for what is known as formalism.

Formalism is a school of thought that advocates the meaning of form over content. It is a doctrine that claims mathematical concepts only have semantic meaning. Mathematics is a constructed set of rules that hold as long as the rules are followed. It can be completely formalized and can determine truth on its own. In the science, the formalism is a method that involves using symbolic language and abstracting information where the meaning of terms are defined as precisely as possible. The involved concepts do not need to be empirically grounded.

A formal system is an axiomatic system. Axioms are formulated in a limited symbolic language, and with the help of specific rules of derivation, certain formulas can be derived, called theorems. A derivation system can determine which inference rules are allowed and the steps of argumentation performed in a derivation. According to Hilbert, the mathematics is a language invented by humans.

Kurt Gödel

Kurt Gödel is an Austrian-American mathematician and logician who had a profound impact on philosophical and scientific thinking in the 20th century. He is known for his incompleteness theorems, which show that Hilbert's attempt to find a complete and consistent axiomatic system for all mathematics is impossible. Mathematics cannot be fully formalized. Formal arithmetic does not prove all true statements.

There are statements that are true but cannot be proven from the given axioms.There is in every formal system a true but undecidable formula that cannot be proven, or whose negation can not be proven. No mathematical formalism is complete. Formal mathematical systems are based on different axioms, and not all formal systems can be proven. We must accept that certain statements are unproven in mathematics. We can be aware of this and still use mathematics. There are mathematical patterns everywhere in nature, while symbols, equations, and conventions are human inventions. According to Gödel, mathematics is a language in nature.

35. SCIENTIFIC PARADIGM SHIFTS

There have been several paradigm shifts and milestones in the history of physics, four of which are attributed to Galileo Galilei (1564-1642), René Descartes (1596-1650), Karl Popper (1902-1994), and Thomas Kuhn (1922-1996). Each of these figures introduced radical changes in perspectives on the nature of science.

Galileo Galilei

Galileo Galilei was an Italian astronomer, physicist, and engineer, known as the father of modern science. For Galilei, nature is a source of knowledge. In nature, there are universal truths that apply at all times, in all places, and for all people. Through investigating nature with experimental methods, it is possible to gain knowledge about fundamental reality. To conduct science, it is important to learn the language of nature, which consists of mathematics. The task of science is to discover the mathematical laws underlying the structure of the universe. Science should observe these laws, and hypotheses about nature should be proven through observation and experiment.

Galilei demonstrated through experiments that the heliocentric worldview described by the Polish astronomer and mathematician Nicolaus Copernicus (1473-1543) was correct. Copernicus described a cosmos consisting of solar systems and galaxies, where the sun (helios) is at the center with planets, including Earth, orbiting around it. This heliocentric model contrasted with the geocentric worldview, which was prevalent during antiquity and the Middle Ages, where Earth was considered the center of the universe with the sun, moon, and planets orbiting around it.

René Descartes

René Descartes was a French philosopher and mathematician. For Descartes, the human intellect is a source of knowledge about reality. A prerequisite for scientific research is the ability to make rational decisions and choices, which requires an intellect capable of doubt. Science should be based on defensible, rational arguments that do not contradict but are integrated with scientific facts. These assumptions have been embedded in natural science since the 17th century. Descartes rejected the medieval view that everything in nature had a soul.

Descartes argued that nature had no soul or intelligence. Animals and plants were soulless machines, just like the human body. Humans are separated from nature by their capacity for intelligence and rational thinking, possessing a soul capable of logical thinking and language. With Descartes, the world became divided into a material world consisting of nature and an immaterial world consisting of the human soul. This led to a sharp duality between body and soul, animals and humans, and science and religion.

Karl Popper

Karl Popper was an Austrian-born British philosopher and science theorist, known as one of the most influential philosophers of science in the 20th century. Popper laid the foundation for the rational empirical knowledge paradigm. The task of science is to distinguish between science and non-science. Scientific theories should be falsifiable through rational thinking and empirical methods. It should be possible to test theories in a way that allows them to be rejected as true or false. From the multitude of possible theories that can explain the phenomena under investigation, the false theories should be eliminated, and the researcher should rationally choose between the remaining ones.

Scientific knowledge is separate from subjective experience and states of consciousness. Subjective knowledge is based on conviction and the pursuit of certainty. For objective knowledge, there is no certainty, only a preliminary hypothesis, and knowledge is hypothetical. A scientific theory can never be definitively proven. It can never be fully verified even if it has withstood rigorous testing over a long period. Even if a theory has a high degree of confirmation, it can only be provisionally retained as the best available theory until it is disproven by a better theory.

Popper had a cumulative view of scientific knowledge development. Scientific knowledge grows through logical steps and adjustments within theories. Researchers' observations are selectively designed to test the extent to which a given theory works as a satisfactory solution to a given problem. It is by engaging with a problem within a theory that a researcher should primarily make their observations.

Thomas Kuhn

Thomas Kuhn was an American professor, science theorist, and author known for the concepts of normal science and paradigm shifts. In his book "The Structure of Scientific Revolutions" (1962), he describes scientific knowledge development as consisting of two phases: normal science and paradigm shift. Normal science is the phase where knowledge develops through logical inferences and small adjustments to already known theories. It is the phase when a researcher adheres to the thought patterns and beliefs prevailing in the given paradigm of the scientific community, characterized by problem-solving within established thought patterns and conceptual frameworks.

A paradigm is a term that denotes the general concepts and thought patterns for how science should be conducted in the scientific community. It determines which questions should be asked, how experiments should be conducted, what results can be expected, and which answers are relevant. It is knowledge considered to be factual. Every scientific community has its paradigms, methods, and specific perceptions of what constitutes validity.

A paradigm shift is the phase where knowledge develops through scientific revolutions. In a paradigm shift, there is a change in the scientific thought patterns and beliefs prevailing in the scientific community. Previous goals and methods for conducting science are perceived as inadequate, misguided, or meaningless. This gives rise to new questions and new experiments to be conducted. Paradigm shifts are irrational in the sense that they cannot be explained based on previous theories and facts. They explain what cannot be explained within the prevailing paradigm and provide a qualitatively different understanding of the fundamental nature of reality. They replace an explanatory system rather than explaining anomalies. An anomaly refers to an empirical prediction that did not turn out as expected.

According to Kuhn, a paradigm shift is preceded by a scientific crisis. The scientific community is governed by paradigms it is not always aware of. Ideas and thought patterns can be so ingrained that anything contradicting the paradigm's validity goes unnoticed, or researchers consider their "truth" superior to anything contradicting it. When a critical mass of anomalies has accumulated, research reaches a critical point where these anomalies must be explained. When someone breaks the prevailing ideas and thought patterns within the paradigm, research enters a pre-scientific phase where the new paradigm is established. The paradigm shift occurs when the new paradigm gains acceptance, followed by a new phase of normal science. Albert Einstein's theory of relativity and quantum physics are examples of the two most recent paradigm shifts in physics are. These are fundamental branches of modern physics that have been revolutionary in the 20th century, and influenced our understanding of fundamental reality but have different views on its nature.

36. RELATIVITY THEORY AND QUANTUM PHYSICS

Einstein's theory of relativity he theory and quantum physics (also called quantum mechanics) are two mathematical models with different focuses on reality. The theory of reativity focuses on the macroscopic world, while quantum physics focuses on the microscopic world. The macroscopic world encompasses the movements of celestial bodies and gravitational effects. For us, the macrocosm includes all the celestial bodies in space: the Earth, the sun, the solar system, galaxies, and others.

The microscopic world involves the movements of atoms and subatomic particles and the very small energy levels involved in these. In the macroscopic world, the physics follow a mechanical and deterministic process. In the microscopic world, physical states must be understood according to a probability distribution.

Einstein's theory of relativity comprises two theories: the special theory of relativity (1905), and the general theory of relativity (1915). These theories have revolutionized our understanding of the concepts of time, space, and the fundamental nature of the universe. The special theory of relativity has changed our view of time and space at high speeds, and the general theory of relativity has provided a new understanding of gravity and its effect on time and space.

The special theory of relativity gives rise to what is called flat spacetime. Flat spacetime is a merging of space and time into a single, interwoven continuum. In the special theory of relativity, the famous equation describing the relationship between energy and mass is derived: E=mc^2. This equation links energy (E) to mass (m) and shows that mass can be converted into energy and energy into mass. The speed of light (c) is constant in a vacuum and is independent of the motion of the light source or the observer. Movement causes time to slow down. A moving clock can show less elapsed time compared to a stationary clock, and an object moving at high speed will experience contraction in the direction of motion.

The general theory of relativity is a geometric theory stating that gravity is not the same force as in Newton's theory of gravity but rather a curvature of spacetime caused by the presence of mass and energy. Objects move along the "straightest" possible path in this curved space, perceived as the attraction of gravity. Space curves around objects with mass because an object with mass creates a gravitational field. Space can both bend and curve depending on how objects with mass move in it. Matter determines the curvature of space, but space and its curvature also determine the movement of matter. Time moves at different speeds depending on the gravitational field.

Einstein's theory of relativity places the foundation of all existence in a block universe. All existence arises from a four-dimensional space filled with geometric structures, whose shapes are determined by immutable laws. Instead of being the three-dimensional space perceived by our senses, the universe is a four-dimensional spacetime, with time added as the fourth dimension. The universe is a static block universe where everything can be predicted, and the laws of physics govern which sequences of events are possible.

Quantum Physics

Quantum physics states that at the foundation of all existence is a vacuum quantum vacuum). It is not an absolutely empty space but contains electromagnetic waves and particles. There is continuous activity in the empty space, where particles and their antiparticles are created from nothing, pop into existence for a moment, and then annihilate with antiparticles. Energy rises and disappears in extremely short time intervals. A syninym for a state of quantum vacuum is the zero-point field.

Zero-point energy is the lowest possible energy that a quantum mechanical physical system has, and it is the energy in the system's ground state. All quantum mechanical systems have a zero-point energy. At the quantum level, particles exhibit both wave and particle properties. This was made particularly evident in an experiment called the "double-slit experiment," where individual electrons show interference, which is a wave property that appears in coopration between two or more wave movements, despite also being detected as individual particles.

Quantum physics accounts for the existence of entanglement. Entanglement is a term meaning that particles can affect each other's states without apparent "causal" interaction. If two particles are entangled, a change in one particle will immediately cause the properties of the other particle to change, regardless of the distance between them. Entangled particles share the same quantum state, but this entanglement is broken as soon as part of the particle pair is measured. Another strange property in quantum physics is that the act of measurement affects the outcome of an experiment. This has sometimes been interpreted as the act of observation collapsing the wave function's possibilities into a single reality.

Unlike classical physics, where the future of a system can, in principle, be predicted exactly if all its present properties are known, quantum physics' predictions are based on probabilities, not exact predictions. There is an inherent uncertainty in quantum physics.

Relativity and quantum physics both make accurate predictions within their respective domains but are incompatible with each other. One reason for their incompatibility is that they treat time differently. It is meaningless to talk about time on small scales in quantum physics, where space is vibrating energy, and things are described differently than with time. In the quantum physics, time is not fundamental but an interaction between particles.

Another significant difference is that relativity is deterministic, while quantum physics is probabilistic and non-deterministic. Relativity is successful in explaining phenomena on large scales, while quantum physics is successful in explaining phenomena on very small scales. Both these theories are milestones in modern physics, but they offer different perspectives on the nature of reality. There have been attempts to integrate these theories into a single coherent theory and remain one of the greatest challenges for physics to solve.

Carlo Rovelli

Carlo Rovelli is an Italian theoretical physicist known for his significant contributions to quantum gravity research. In his book "*Helgoland: The Relational Interpretation of Quantum Physics*" (2022), he presents his revolutionary interpretation of quantum physics. He describes a reality fundamentally composed of processes, events, and relationships. This theory focuses on interactions rather than isolated objects, asserting that to understand nature, we must examine interactions. Without interactions, there would be no effects on anything; nothing could attract or repel anything.

There are no physical properties like speed and position in isolation. Speed is a property an object has relative to another object. A stone has no position in itself; it only has a position in relation to another stone. The sky has no inherent color; color is related to the eye and has no definite position. Up and down are not absolute concepts but are relative to our position.

An object, taken by itself and independent of interaction, has no specific state. At best, we can assign it a kind of probabilistic disposition to manifest itself in one way or another, but this is always relative to another object. Everything is in relation to something else. We understand things only by relating them to something else. Relational thinking can be found in all sciences. In biology, living systems are characterized by being understandable in relation to their environment, shaped by other living beings. In psychology, personality exists in a relational context. In chemistry, the properties of elements are determined by how they interact with other elements. In many other fields, we understand things through their existence in relation to other things.

Reality does not consist of independent entities with determined properties but of an extensive web of interacting entities manifesting through their interactions. In this web, entities are nothing but short-lived nodes. Their properties are not determined until the moment of their interactions. Properties are the ways they affect and act on other objects. They assume value when they interact, and this value is determined in relation to the objects in interaction. They have properties only in relation to others and only when they interact. There is no way to view reality that is not dependent on perspective. Everything is what it is only in relation to something else.

The world is not as solid as we are accustomed believe in our macroscopic world. In the microscopic world, there is no permanence or solidity. The quantum world is made of discontinuous events that are probable and relative, without duration. Every interaction is an event, and these events weave reality, not objects with absolute properties. The theory goes beyond the idea that the world is made of substances with properties and forces us to think of everything as relations. It opens up a new perspective on reality, composed of relations rather than objects.

After formulating his theory of the relational nature of quantum physics, Rovelli began reading texts on emptiness written by the Indian philosopher Nagarjuna, who lived during the second century CE and was one of the most prominent and influential philosophers and thinkers in the Buddhist tradition. Rovelli found it remarkable that Nagarjuna, even in his time, had thoughts that resembled his interpretation of quantum physics.

Nagarjuna's Philosophy of Emptiness

Nagarjuna is considered the founder of a philosophy known for its emphasis on emptiness, which means that all phenomena are empty of autonomous existence or essence. This means that objects only exist in relation to other objects. All existence and all properties of objects arise through their relations and interactions with other objects. According to Nagarjuna, all structures are empty. They exist only when we think of organizing something else. They are neither precursors nor non-precursors of objects. Stars do not exist in themselves but arise in the encounter with sensory impressions and the thoughts they evoke in our brains. There is no self that observes the star. The self, articulated by language and thoughts, has no inherent reality. It is an interconnected multitude of phenomena that constitute it, and each phenomenon depends on something else. Everything comes from the encounter with other things.

We live in a phenomenal world of mutual dependence and contingencies, says Nagarjuna. It is a conventional world we experience with an illusory perspective. It is akin to a dream and more like a projection of human consciousness. This means that experiences of phenomena are not real in the sense that they do not have their own nature. They depend on something else and lack enduring substance. Because everything is about relationships, it is pointless to seek an ultimate reality. There is no ultimate substance, no starting point from which everything follows, and no primary substance on which everything depends. It is not reasonable to seek an ultimate reality, but this does not exclude investigating the phenomenal world to understand it. We can still explore and analyze the world and reduce it, but it is not a world to be derived from something absolute.

Western philosophy has long tried to answer the question of what fundamental reality consists of, says Rovelli. Science has sought the smallest building block of the universe that can explain the origin of everything and give rise to everything else. Physics has sought a primary substance that is the starting point from which everything follows and on which everything can depend. First, it was the atom, then the atomic nucleus, and then smaller particles. Over time, it has become clear to science that these "particles" are nothing more than patterns of vibrating energy. Today, the foundation of reality is described as a quantum field defined mathematically.

Physics has provided a picture of reality as a large space where particles move according to fundamental laws. Particles follow these laws automatically, without anything happening. When particles collide, something happens. Particles are real and move according to fundamental laws, but physics does not know where these laws come from. In quantum physics, things happen differently.

Particles are not real, and they do not move according to fundamental laws. Spacetime is not fundamental in quantum physics. Spacetime is constructed from discrete phenomena that build up the continuity of space and time, but space and time do not exist from the beginning. Spacetime arises from something else that we do not understand. It is not evident that matter is the building block on which the universe is constructed.

Quantum physics has destroyed the picture of reality as made of particles moving along determined paths. Reality is fundamentally uncertain, and the mathematics of quantum physics does not give us the reality we experience.

Scientific thinking is not just about achieving certainty. Scientific thinking involves finding new ways to conceptualize the world, and sometimes it leads to something radically new. Scientific theories provide us with a vision of reality, a conceptual framework, but they only work for certain environments where everything behaves the same way. Science describes everything in terms of something else. Our perception of the world is very limited. We need to question our concepts and be prepared for rethinking. This aspiration makes scientific thinking so effective, says Rovelli.

The task of science is not only about making predictions. The task of science is also to explore new ways of imagining, understanding, and gaining insights into the nature of the world. One of the biggest mistakes humans make is to seek certainty in their attempts to understand everything. New knowledge is not generated by certainty but by the absence of certainty. This means we should be open to doubt and aware of our ignorance to understand better. New understanding comes through abandoning assumptions that seem self-evident.

Rovelli argues that the discovery of the quantum world is too radical not to be relevant to questions concerning the perception of the material world, consciousness, and the sense of self. Quantum physics affects microscopic structures that give rise to our bodies but does not help us understand our subjective experiences and sense of self, which involve the brain on a macroscopic level. The question is how consciousness, the sense of self, beauty, and meaning can arise from matter. Quantum physics contradicts the idea of matter as a universal substance defined by mass and motion. The sense of self is linked to an integrated set of mental processes and emotions. We take the process to be a unified entity, but even that entity is relative to something else.

We are part of nature among countless other natural phenomena. Our mental life requires the existence of neurons, sensory organs, a body, and the complex processing of information that occurs in our brains. Everything suggests that without this, we have no mental life. The world is better described by relative variables and their correlations. This allows us to be freed from the dichotomy between objective matter and mental life. The rigid distinction between a mental and a physical world can be dissolved. We can think of both mental and physical phenomena as natural phenomena generated by complex structures in interaction. Both involve interaction with the physical world.

Lisa Feldman Barrett

Lisa Feldman Barrett is an American neuroscientist and professor of psychology, known for her revolutionary views on the relationship between emotions and the brain. In her books "*How Emotions Are Made: The Secret Life of the Brain*" (2017) and "*Seven and a Half Lessons About the Brain*" (2020), she presents her theory on emotions and the brain. Central to her work is the idea that the brain and body together create events that are experienced as emotions. An emotion occurs when the brain makes something meaningful out of sensory inputs from the body in relation to the world.

Barrett asserts that what happens in construction of emotions is a prediction of what should be done. The brain is wired to send these predictions to sensory systems, which anticipate our emotions. We do not feel with our sensory organs but with our brain. What we see is a combination of what is outside and what is in our brain, and similarly, our brain constructs what we feel inside our body. The brain also adds information from previous experiences to guess what these feelings mean. It may seem like we feel first and act later, but the feeling actually comes second in our brain.

Many researchers believe that emotions come from specific parts of the brain, triggered by the world around us. They rely on the assumption that emotions are built into our brains and bodies. This understanding of emotions is incorrect, says Barrett; emotions are not triggered, but we create them. Emotions are constructed, and there are no traces of them in the brain. Our emotions arise as a combination of our body's physical properties and a flexible brain that adapts to the environment it evolved in, which includes our culture and upbringing.

There are no universal emotions; they vary from culture to culture, and there are no pre-programmed emotions in our brain and body. They are psychological experiences that each of us constructs from our unique personal history, physiology, and environment. Emotions are a variety of events where our brain tries to make what happens most functional in a specific situation. An emotion includes a whole group of events that can be expressed in different ways, depending on context, all with their specific expression in body movement.

In neuropsychology, it is common to define emotions with specific facial expressions and body movements. The same facial expression and body movement can be shown for a particular group of emotions. Many have taken this as evidence that there are universal expressions for emotions. However, recent data has shown that there are no universal expressions for emotions. Our brain usually has several ways to handle a particular situation. The brain creates a cascade of predictions and estimates probabilities for each of them. If our brain has predicted well, our neurons have already fired in a pattern that matches incoming data. This means that sensory data itself is of no use except to confirm these predictions. The brain is not focused on accuracy but on keeping us alive. The brain creates our reality.

There are no universal mental properties, but they are still useful. A mood is a useful property. It is a general feeling that comes from our body, called affect. Our brain produces affects all the time, whether we notice them or not. Our brain summarizes what is happening in our body at the moment, and we feel that summary as affect. The affect must be made meaningful for our brain to execute a specific action. The construction of emotion is a way to make an affect meaningful. Affects do not explain how we construct our emotions or what feelings mean, what we should do with them. Feelings of affect can be pleasant or unpleasant, and active or inactive. Affects are real, and emotions are emergent.

Our ability to predict and use energy efficiently is key to survival. The most significant work of our brain is to predict the body's need for energy before it arises so that we can make valuable movements and survive. We can think of energy efficiency analogously to a financial budget. It is about knowing when to spend economic resources and when to save them. The best way to maintain a financial budget is to be sure that we have resources and can anticipate financial opportunities before they arise. Similarly, a body budget is required to maintain a body. The body's primary function is to allocate energy resources to the body's internal systems and keep the body's energy system in balance.

Affects are a summary made by the brain about the state of our body. They are an indication of whether our body's budget is in balance or not and help regulate the body budget. The brain's primary task is to regulate the systems in our body: the respiratory system, the cardiovascular system, the endocrine system, the immune system, metabolism, and more.

In every moment, decisions are made about which energy should be spent or saved. Every activity is a choice where our brain tries to predict when resources should be spent or saved. Our brain prepares for action and makes changes in the body budget. The physiology of the internal systems prepares the body for action. It is not about thinking and making decisions consciously by weighing pros and cons. It is about something happening inside us to predict a certain amount of movements rather than others. This reflects a determination of value, which is the value that each movement is intimately linked to the body budget through allostasis.

Allostasis is defined as the process that maintains homeostasis through adaptive change in the body's internal environment to meet perceived and anticipated demands. The brain is constantly trying to keep various systems in balance. If it is hot, we start to sweat, and if we are threatened, our pulse increases. This process is called allostasis, and it means that the brain resolves a deviation by adjusting something else. In allostasis, deviations from the normal occur, in terms of, for example, hormone levels, blood pressure, and breathing rate. The purpose is to adapt the body to some load in the environment. Homeostasis is about the body's ability to maintain a certain stable state despite external changes. It maintains balance by changing physiological parameters. If homeostasis waits for deviations and then corrects them, allostasis uses prior knowledge, both innate and learned, to prevent deviations and minimize them.

Our brain seeks to keep various systems in balance by predicting the energy needed before movements arise, so that we can survive and make valuable movements efficiently. This means that it predicts and prepares to meet the body's needs before they arise. The best source of information is that which comes from past actions taken at other times under similar circumstances. If a previous action has been successful, it is more likely to be repeated. All kinds of animals, including humans, prepare their bodies to act through past experiences, says Feldman.

The brain predicts actions based on past experience. Past experiences include not only what has happened in the world around us but also what happens inside our body. Our brain is set to initiate actions before we are aware of them. It launches our next set of actions based on past experience and the current situation, doing so without our awareness. Our actions are under the control of our memory and our environment. Our actions today become our brain's prediction of actions tomorrow. We cannot change the past, but we can invest time and energy to learn new things and try new activities.

We can change how the brain predicts the future. Through practice, we can make automatic behaviors more likely than others and have more control over our future actions and experiences. There is some freedom to refine our predictions in new directions, and we have some responsibility for the outcome. By owning a predictive brain, we have more control over our actions, more experience, and more responsibility than we might have thought or wished. We are not separate individual entities but have a socially dependent nervous system, and how we treat each other matters in a very tangible way. We influence others' brains and bodies with our actions and words.

According to Feldman, it is wrong to assume that the brain evolved for thinking, where the human brain is considered the most sophisticated thinking brain. There is no scientific support or evidence for this. Neurobiological studies in recent decades have shown that the brain did not develop for thinking. A common assumption is that the brain evolved in a progression from lower animals to higher animals. This model describes the brain as consisting of three separate parts and is called the triune brain. It aims to explain how evolution has affected the brain's division into three parts.

The American neuroscientist and physicist Paul MacLean introduced the triune brain in the 1960s as a model for brain development. He claimed that the further out a part of the brain is, the more advanced it is. After serving as a model for brain development for a long time, MacLean's model is now considered outdated (Patton 2021).

The triune brain describes the brain in three separate parts, consisting of the reptilian brain, the limbic system, and the neocortex. The reptilian brain includes the brainstem and the cerebellum, which is located beneath the occipital lobes of the cerebrum. It is called the reptilian brain because a reptile's brain is largely made up of the brainstem and the cerebellum. From here, our movements, our balance, and our ability to act in relation to the environment are coordinated. It controls muscles, autonomous functions, such as the unconscious control of the heart, lungs, stomach, liver, and balance. It takes care of our primary instincts, and is the seat for survival instincts such as the freeze response, fight or flight response, and for affective systems. It has two hemispheres and three lobes. The anterior and posterior lobes are responsible for the unconscious parts of muscle movements, while the third lobe is responsible for balance.

The limbic system encompasses several groups of nerve cells that mediate feelings of anxiety, aggressiveness, and more. It is referred to as the 'emotional brain,' and this is where the neural circuits that handle our affects and subconscious thoughts are located. It plays a crucial role in our ability to experience and process emotions, as well as in learning, the ability to form memories, and our motivation. This is also where our sense of smell center and the structure known as the amygdala are located, where sensory impressions are linked to a feeling. It has also been called the 'mammalian brain' because it was presumed to be inherited from ancient mammals. The neocortex is the outermost part of the cerebral cortex in the brain. Its name derives from being the evolutionarily newest part of the cerebral cortex. It consists of six layers of neurons that run along the surface of the brain. This part of the brain houses higher developed behaviors and cognitive abilities. It perceives and interprets all our senses, and also includes speech, logic, abstract and rational thought, which are considered unique to humans

According to the triune brain model, the brain developed in layers that allowed rational thinking to overcome our animalistic desires and emotions. It is an idea of a struggle between rational thinking and primitive animal emotions and survival instincts. It is a myth that natural selection is directed specifically toward us, says Feldman. We have only specific adaptations that have helped us survive and reproduce. Animals are not inferior to us. Even animals are effectively adapted to their environment.

Since Plato, the brain has been described as a struggle between rational thinking and primitive emotions and survival instincts. The Platonic idea of the struggle between rationality and emotions has long been an explanation for human behavior in Western culture. Thinking is considered rational, and emotions are considered irrational. To move beyond the idea of the struggle between rationality and emotions, we need a fundamental rethinking of what it means to be rational.

Feldman questions why the brain has evolved to function the way it does. It is not a question that can be answered because evolution does not act based on goals, but we can say that the brain's most important work is not rational thinking. The brain's primary function is to allocate energy resources to the body's internal systems, not primarily for rational thinking.

Our brain is a complex network organization consisting of 128 billion neurons connected to a single, massive, and flexible structure. Each neuron looks like a small tree, with branches at the top, a long trunk, and roots at the bottom. The bushy branches, called dendrites, receive signals from other neurons, and the trunk, called an axon, sends signals to other neurons through its roots. When a neuron fires, an electrical signal rushes down its trunk to its roots. This electrical signal causes the roots to release chemicals into the gap between neurons, called synapses. Chemicals are then transferred into the synapses and attach to another neuron's branches, causing that neuron to fire. In this way, information is sent from one neuron to another neuron.

The arrangement of dendrites, axons, and synapses connects all the brain's neurons into a single network. Being complex is the brain's ability to create a diversity of distinct neurological patterns of activity. The brain's network is not static but always active and continuously changing. In some parts of the brain, neurons die, and in others, neurons are born. Connections between neurons become stronger when neurons fire together and weaker when they do not.

The brain's neurons always communicate with some other neurons, but not all communicate with all. All are engaged with what is nearby. This strengthens pathways in the brain and makes abstraction possible. Abstraction allows our complex brain to make flexible predictions based on function rather than physical form. When information from the external world enters the brain, some neurons fire more frequently than others. This causes gradual changes in the brain, called plasticity, and drives the brain's development towards higher complexity via two processes. The first involves strengthening connections between neurons, especially connections that are used frequently. The second involves weakening less-used connections, and those we do not use are lost. Microscopic parts of neurons change every day gradually through these processes. For example, the associated nerve connections become more efficient, and dendrites more bushy.

Humans' brains have the same type of brain structure, yet each individual brain has its own mental reality. The mental and the body are strongly connected, and the boundary between them is porous. Our brain prepares our body to act, thus contributing to mental experiences. We perceive the world differently depending on our predictions. A particular human brain in a particular body, arising and activated in a particular culture, will produce a particular type of human mind. The culture we live in shapes our brain. The significance of thoughts and emotions varies across cultures. In Western cultures, thoughts and emotions are perceived as fundamentally different from each other, while in Eastern cultures, thoughts and emotions are interconnected.

We all live in a world of social realities. Our self is not objectively real by nature but depends on social interaction with other people. Social psychology says that there are multiple selves. The fact that our brain can keep track of all the varying selves, such as being a child, teenager, adult, and elder, is because we always have a body. The self can be constructed in the same way that predictive systems construct emotions when categorizing the continuous stream of our sensations from the body and the world, says Feldman.

37. TIME AND CAUSALITY

It is not only the perception of reality, consciousness, and the concept of self that need reconsideration, but also the concepts of time and causality. A debated question is whether the future is predetermined and if time is absolute or relative. This question touches on subjects related to determinism, free will, and the direction of time.

All living beings have an internal biological clock that

regulates our circadian rhythm and other life processes. This sense of time is crucial for our survival. Our perception of time varies. Subjectively, we can experience time standing still, other times it can feel like time flies by, or that it passes very slowly. Time can be perceived differently depending on the culture we live in. Some cultures value punctuality and strict schedules, while others have a more relaxed approach to time. For some, time is a cyclical process, while for others it is a linear path from creation to end. Since ancient times, humans have tried to periodize and measure time in various ways. Time has been described as moving in cycles in nature, such as day and night, different seasons, and ages.

In ancient Greece, the Greeks imagined time in different ages, which ended and were created anew. Similar thoughts about cyclical time are also described in Hinduism and Buddhism, while Islam, Christianity, and Judaism view time as linear, without a beginning or an end. There are many attempts to explain time, but no clear definition of time. Time has been explained as a flow moving forward, an expansion of the universe, as in cosmology, and a dimension intertwined with a three-dimensional space. Time is not fundamental in the Wheeler-DeWitt equation, which describes mathematical wave function of the entire universe, where neither spatial nor temporal aspects are included. Time has been described both as subjective and objective.

Subjectively, time refers to the inner mental world. It is a perception of time that cannot be measured objectively. Time is a category in our perception and inherent in our thinking to comprehend the world. It is our mind that creates an illusion of time, and there is no flow, just a sequence of mental states. Time is a sequence of events. If time is a sequence of events arranged chronologically, then maximal time is when nothing happens. There is no experience that does not happen now. The past is a memory stored from a previous now, which arises in the present. The future is a projection of our intellect, a fantasy that arises in the present. It is in the present moment that we can plan and work to achieve goals, and where our lessons from the past become relevant and applicable.

Time is a scarce resource in our society. We are often so busy with everything that needs to be done that the present becomes merely a means to achieve our goals. We perceive the past and future as relevant and meaningful, and the present as irrelevant and meaningless. We can take a reverse perspective and allow the present to be relevant and meaningful.

Our perception of time varies. Subjectively, we can experience time standing still; other times, it can feel like time flies by or moves very slowly. In our everyday life, we experience time as a flow in one direction. A metaphor is to see time as a long line with a series of events occurring on the timeline, where the interval between two points on the line is measured with clocks. This is a way of thinking about time that works well in our daily life, but it is not true for the universe as a whole.

Objectively, time refers to the external physical world. It is a perception of time that can be measured and is a measure of the duration of the time consumed. Time regularly measures recurring events. It can be what it takes for the Earth to rotate around its axis or the clock that shows the time of day. All clock time is essential for societal and social interaction between people. Clock time is needed to organize, predict, and plan events in social life. Society could not function without precise timing, nor could research, especially research in astronomy.

Time was perceived as a fixed and unchangeable quantity until Einstein's presentation of the theory of relativity. He defines time as the distance between events along the fourth axis of spacetime. According to Einstein, time is affected by large masses and their movements, which causes time to run at different speeds in different parts of the universe. If we take two clocks that show the same time and lift one clock up and down after a few seconds, the lifted clock will show that time has gone faster, and the clock below will show that time has gone slower. This is because a large mass slows down time and makes the clock below tick slower. The difference is so small that it is not noticeable.

One reason quantum physics and the theory of relativity are incompatible is that they treat time differently. On the small scale in quantum physics, the concept of time does not work. It is meaningless to talk about time where space is vibrating energy. Things must be described in a different way than with time. In quantum physics, time is an interaction between particles. It is possible to manage without time and only talk about how something moves. Time is relative to movement, and movement is relative to time. Time can be measured as movement, but movement can also be measured as time. This means that time is not fundamental, but it is useful for organizing our daily experiences. Events are fundamental and follow one another, and we interpret this as time. It may be more accurate to say that the flow consists of these things happening. Time is derived from a sequence of events. It is not the flow but the events that are important.

According to modern physics, there is no indication of the direction in which time flows. The words "past" and "future" lose their meaning because time has no particular direction. There is nothing that distinguishes between the past and the future and no moment that is more special than another.

Modern physics' conception of time contradicts the way we perceive the world. Normally, we experience the past as something entirely different from the future. We know we get older and not younger. We imagine that time has a direction, and we can think of something we remember in the past or something that will happen in the future. We experience time as something we move through, and that time is a movement separate from our movement in space. We can move forward and backward in space, but not forward and backward in time. Unlike movement in space, movement in time happens automatically, whether we want it to or not, and it can only go in one direction.

We perceive that there is a difference between the past and the future that we call a "now." From a subjective perspective, there is a special moment that we experience as more special than any other, which is the present. If everything is predetermined, and there is no "now" in a temporal sense, it means we do not have free will. This contradicts the experience that we can choose between different possible actions.

The question is whether time has a direction or is a conceptual misunderstanding. There is no mathematical description that explains what the passage of time really is. It is difficult to find an explanation for the relentless flow of time in one direction. Entropy is an attempt to explain the passage of time och is a measure of disorder. The higher the disorder, the higher the entropy. Entropy was established as an explanation for the passage of time in the second law of thermodynamics. According to this law, entropy in a closed system tends to increase and cannot decrease. Entropy, in a closed system grows spontaneously over time without external influence.

There is also a first law of thermodynamics. According to this law, energy can neither be created nor destroyed; it can only be transformed. These two laws were considered fundamental. They cannot be reduced to or explained by anything more fundamental. The constant increase in the system's disorder has been taken as an explanation for time's asymmetry between the past and the future. The view that time is a progression from order to disorder requires that time has a beginning. The question is whether time has a beginning and an end. Some argue that time has neither a beginning nor an end. What is called the "big bang" is not the starting point for time but a transitional stage in an eternal universe.

Roger Penrose

The British mathematician and Nobel laureate in physics (2020) Roger Penrose argues that time has a beginning and an end. In the distant future, when all mass has been consumed by black holes, the universe will lose temporal dimension. When there is no mass, there is nothing to measure time with. If we remove mass from physical equations, it is no longer meaningful to talk about time. When all mass in the universe has been consumed by black holes, evaporated, and left only radiation, the situation is the same as at the birth of the universe. The dying universe resets time, and the universe will be reborn in a new "big bang." According to Penrose, "an old big universe" is fundamentally identical to "a new young universe." An old and a new universe are fundamentally the same, and the dying universe will self-generate.

Richard A. Muller

Richard A. Muller is an American emeritus professor of physics. In his book "*The Physics of Time*" (2016), he writes that Einstein's theory of relativity has not succeeded in abolishing the objective present. In the theory of relativity, space and time are so closely connected that time grows with space. He reintroduces the objective present and argues that there is a now that distinguishes between the past and the future. Muller refers to the constant expansion of the universe, that it is the expansion of space that creates more time and is the present. The expansion of spacetime is analogous to what happens when we throw a stone into a lake. The stone creates rings on the water in the form of waves, which increase the surface area.

According to Muller, the past exists in spacetime, while the present is continuously created due to the expansion of spacetime. We do not move to an already existing future because the future consists of a long series of "now moments" that have not yet been created. The past exists in spacetime, and the future is not determined. Thus, the theory allows free will. Future events are determined by the decisions we make in an endless series of "now moments" that constantly add new slices of spacetime to history.

Muller argues that his theory can be supported by studying the gravitational waves that occur when two black holes collide. When they collide and merge, gravitational waves are created that propagate in space. It is possible to observe patterns of gravitational waves long after the black holes have merged. The waves create patterns similar to what we see when raindrops fall and create rings on a lake. Even when rain has stopped, we see ripples on the lake's surface. If the black holes that merge are large enough, it is possible to detect and measure gravitational waves.

In 2015, vibrations from two black holes merging far away in the universe were recorded. That event lasted only a fraction of a second and resulted in a black hole of 62 solar masses, an eruption of radiation, and gravitational waves that propagated through space. It was LIGO detectors that measured these waves as vibrations in spacetime. The vibrations were so weak that the detectors' mirrors could only move by the equivalent of an atom's diameter. The measurements were too uncertain to determine Muller's expectation that the expansion would create a time delay in the final gravitational wave compared to the rhythm measured in the first waves.

Causality

The relationship between causality and time is deeply rooted in our perception of how things happen in our everyday understanding of the world. Causality is a concept that refers to the relationship between cause and effect. When there is causality between two events, one is called the cause and the other the effect. The starting point is that an event or action directly results in or brings about another event. Causality is central to scientific, philosophical, and everyday reasoning, as these often involve seeking causes for events and making predictions based on them.

The problem with causality is that the first cause can never be found. In thermodynamics, the so-called "arrow of time" (direction of increasing entropy) points in the same direction as the causal arrow, meaning that cause and effect move from earlier to later times. Just because two events correlate in time does not necessarily mean that one causes the other. Distinguishing between correlation and causation is central to scientific research.

Einstein's theories of relativity complicate the traditional understanding of time and causality. According to Einstein's theory, two events that are simultaneous for one observer may not be simultaneous for another observer, depending on their relative motions. Quantum physics also introduces more intricate causal relationships, especially with phenomena such as entanglement, where particles can affect each other's states without any apparent "causal" interaction. Entanglement implies that if particles are entangled, a change in one particle will immediately result in a change in the other particle's properties, regardless of the distance between them.

Immanuel Kant

The German philosopher Immanuel Kant argued that causality is an expression of how the human mind is constructed. There is an inherent pre-understanding in the way we experience the world. Our perception of the world depends on internal mental functions and processes. We would not perceive the world as an ordered, rule-governed place if time, space, and causality were not crucial functions. All knowledge we can attain is a structure that our mind imposes on representations and cannot describe the world as it actually is, independent of our senses.

David Hume

The Scottish philosopher and historian David Hume argued that causality is something we have learned by observing what happens. According to Hume, our expectation that events will unfold in a certain way is a habit. We cannot know what causes what. We see what happens, but not that one event causes another to occur. We only see that it happens repeatedly. There is no reason for events to continue to develop in the order we are used to seeing them. The necessity lies within ourselves rather than in the events.

Things do not have to happen. Causality does not strike the eye in the same way as sensory impressions. It is our intellect that imposes its way of thinking or habits on the world. Our intellect seeks patterns because it is constructed that way. This shifts causality from a category in the real world back to our intellect. We try to place things into general patterns, but we do not know what causes the patterns. It is a goal that we can never achieve. There is a difference between the physical world and the mental world regarding causality.

According to Hume, causality is not the cause of natural laws. A physical law of motion does not say anything about what causes the motion. Measurements reveal a pattern, but they do not provide an explanation for the pattern. Physical laws cannot be intellectually analyzed into something else and are not caused by anything else. Mental events are not causal. They can be intellectually analyzed into something else and have different causes. In physics, causality is a necessity in the relationship between events, expressed in natural laws.

A natural law is a mathematical equation that describes regularities in natural processes. There are regular events in the universe, but they are not the same as laws. Unlike causal laws, regular events in the universe are real. It is not possible to analyze the cause of laws in the universe, but it is possible to analyze the universe in terms of causal laws. Most laws are empirically tested and understood based on deeper laws. They are purely empirical statistical relationships and lack theoretical explanation.

Ulf Danielsson

The Swedish professor of theoretical physics Ulf Danielsson writes in his book "*The World Itself*" (2020) that there are no real natural laws. They are nothing more than representations of what our models of the world have led to so far. They are limited and constantly evolving. There may be phenomena that are forever beyond what we can see. The universe is not governed by natural laws because these are constructed by humans. Natural laws can evolve because they only exist in our minds. They are deterministic in the sense that they predict things.

Determinism means that everything in the universe is predetermined, including the choices we make. There is a conflict between our intuitive sense of having free will and the deterministic laws that determine our choices. An action cannot be free if the world is deterministic and everything is a consequence of the laws of physics. The problem of free will is based on a worldview where the boundary between models and the world is not sufficiently clear. We draw conclusions about the world based on models that are neither nor can ever be complete.

There is a paradox with free will and determinism. Free will should not be traceable to something further that determines it, and determinism's causal chain should be traceable indefinitely backward without finding a single weak link. From that perspective, both free will and determinism are based on the unattainable perspective of omniscience and equally impossible, says Danielsson.

Sean M. Carroll

The American physicist and philosopher Sean M. Carroll argues that the physical level and the mental level are two levels of reality that use different concepts in describing events in the world but are compatible. Reality is described both on a microscopic and macroscopic level. On the microscopic level, free will does not exist, but on the macroscopic level, the concept of free will is useful and real. Macroscopic events depend on interaction with the environment, and its laws cannot be derived from any lower level. Events look different on the microscopic level and the macroscopic level. Both the microscopic level and the macroscopic level involve uncertainties in causality.

In physics, uncertainty is handled by calculating probability and mathematical statistics. Probability calculations make it possible to predict that a certain event will occur and provide a measure of how often a specific event occurs. Mathematical randomness follows the rule that the probability of all possible events is equal. Randomness is what deviates from an expected value.

In physics and mathematics, randomness is a way of explaining processes that cannot be predicted or calculated. Randomness determines what happens when a process is not deterministic. The word deterministic means that future behavior can be predicted using a set of known parameters. Physics distinguishes between deterministic randomness and indeterministic randomness. Deterministic randomness means that there are computational difficulties in analyzing causal relationships, while indeterministic randomness means that there are deviations and uncertainties built into natural laws. There is complexity and chaos in the physical nature that cannot be calculated or explained by natural laws. It is impossible to make deterministic predictions for chaotic and complex systems. Not all events in nature can be explained by referring to causality or some natural law. There are events that cannot be explained by mathematical randomness.

Carl Gustav Jung

In the 1920s, the Swiss psychiatrist Carl Gustav Jung introduced the concept of synchronicity to describe events that coincide in a way that goes beyond causal and mathematical randomness. Jung defines synchronicity as a meaningful coincidence between a mental and a physical event, which share the same or similar significance without known causality. He posits that there are connections beyond those based on cause and effect, and that the inner and outer worlds are interconnected in a deeper, meaningful way.

Jung describes synchronicity as a complement to the classical view of causality in his book "*Synchronicity*" (1998). If the relationship between cause and effect is only statistically valid, then the principle of causality is only relatively useful in explaining natural processes.

This means that the connection between events can be something else than causal under certain circumstances. Viewing synchronicity as an alternative explanation to causality does not oppose causality but complements it, reminding us that science does not have all the answers. The scientific worldview misses aspects that cannot be grasped statistically. The goal of physics is to establish regular events that can be repeated, excluding rare and unique events. It forces nature to answer questions designed by humans. Every answer from nature is influenced by the type of questions asked.

Synchronistic events are often experienced with a sense that there are meaningful connections between various events without apparent explanation. It is an experience of meaning on a personal or symbolic level, related to something beyond the personal.

Jung introduces the concept of the collective unconscious as part of his theories on the psychology of individuals. The collective unconscious refers to a part of the unconscious mind shared by all humans, containing memories and behavioral patterns inherited from our ancestors, which he calls archetypal images.

Jung deeply explored the boundaries of the human psyche and was open to the idea that there are aspects of the psychic and spiritual life that we do not yet understand or that are beyond our current scientific comprehension. There are unknown realms or aspects of the human mind and the universe that we may become aware of in the future, even though they are outside the scope of what he specifically termed the collective unconscious.

The collective unconscious is a network of relationships that transcend space, time, and causality. At the level of the collective unconscious, we are open to influences from the universe and particularly open to synchronistic coincidences in times of change, where meaningful events carry messages that lead to deeper insights and understanding.

The collective unconscious is a set of universal archetypes, which are inherited human characteristics and come from many repeated life experiences. These represent universal patterns of behavior and are inherited in the same way as instinctive behavior patterns. We are born with a complete set of archetypal potential that affects us as deeply as anything else. They are a kind of prototype, pattern, or model of ideas, ways of thinking, and feeling that influence how we experience and interpret phenomena. They are preferred ways in which the psyche manifests itself and a template for a specific pattern of behavior to manifest.

Archetypes are symbolic manifestations of the collective unconscious. They can manifest as images in the dream state and also in the waking state. They are autonomous images that manifest with their own wills and over which we have no control.

Archetypes are themselves neutral but can be interpreted in positive, negative, or neutral ways. Each archetype has its own meaning and symbolizes fundamental human motivations. Archetypes are not fixed and complete in the psyche but are activated when we undergo periods of change and transitions in our lives. These transitions can occur at birth, childhood, adolescence, and adulthood. In transitions a primordial energy is activated that acts as a catalyst to transform the individual psyche and releases the resources needed to meet new challenges in lif.

The collective unconscious cannot be expressed without becoming subjectively conscious. We can only approach it by making it something subjectively meaningful. The collective unconscious is universal and connected to the unconscious of other people. It is a shared unconscious that exists in every individual, group, society, and culture, and is not limited by conditional ways of perceiving and thinking. Beyond the collective unconscious lies the unknown, which we cannot know, do not yet know, but may be possible to know.

The collective unconscious is a kind of memory bank of human experience. It contains recurring themes in ancient myths from around the world and throughout all times. These are myths about gods and heroes that originate from and interact with the mind. Jung's studies of mythological symbolism in different cultures led to his discovery of archetypes. In these myths, he found similar patterns and symbolism.

Jung sought support from physicists for his idea of synchronicity and initiated an interdisciplinary discussion with the Austrian physicist Wolfgang Pauli (1900-1958), known for his Nobel Prize in 1945. Together, they discussed whether a synchronistic event could be a sign that there is an underlying non-causal principle behind subjective and objective realities. The question was whether the organizing forces that shape nature could also be responsible for shaping the psychic structure. They attempted to find a unifying theory that could bridge the gap between psyche and matter

PART IV

SCIENTIFIC INTUITION

In this section, I address the concept of intuition in relation to natural science and present two qualitative studies on scientific intuition: one based on statements about scientific intuition made in a group discussion broadcast on Swedish television, and the other based on statements about scientific intuition made in individual interviews.

38. INTUITION AND NATURAL SCIENCE

The role of intuition in natural science is debated. Traditionally, intuition is considered irrelevant to the development of scientific knowledge. Intuition is knowledge associated with states of consciousness and is therefore deemed to lack value for natural science. The foundation of natural science is supposed to be rationally developed theories that can be empirically tested. Natural science should be free from intuition, beliefs, poitical, philosophical, or economic interests or opinions (Graziano & Ranlin 1997). Through the development of expert knowledge and scientific methods, the significance of intuition should be eliminated. Despite intuition supposedly conflicting with the natural scientific paradigm, several scientists have recounted how intuition has played a key role in their scientific discoveries. Three examples are the accounts given by August Kekulé, Jules Henri Poincaré, and Albert Einstein.

August Kekulé

August Kekulé (1829-1896) was a German chemist known for his significant contributions to the understanding of organic chemistry, particularly molecular structures. One of his most famous contributions was his proposal of the ring-like structure of the benzene molecule. Chemists knew that the benzene molecule consisted of six carbon atoms and six hydrogen atoms, but they could not understand how the atoms were connected.

Kekulé describes two dreams that contributed to his understanding of molecular structures in organic compounds, including benzene. The intuitive insights that came through his dreams played a crucial role in solving these chemical puzzles. Kekulé describes that while pondering the structure of organic compounds, he fell asleep and dreamt of atoms dancing before his eyes. In the dream, he saw the atoms forming chains, spinning around, and intertwining like snakes. Suddenly, one of the snakes caught its own tail and formed a ring. This dream gave him the idea that organic molecules might have a ring structure.

After trying to understand the benzene structure for a while, he dreamt of a circle of snakes swallowing each other's tails. This dream led him to the insight that the benzene molecule might be a ring of six carbon atoms. These dreams, or rather the intuition they provided, were validated experimentally.

Jules Henri Poincaré

Henri Poincaré (1854-1912) was one of France's greatest mathematicians, theoretical physicists, and scientific theorists. Often described as the last universalist, capable of understanding and contributing to virtually all areas of mathematics, Poincaré is known for his significant contributions to mathematics, mathematical physics, and astronomy, as well as his particular views on intuition within mathematics and science.

Poincaré viewed intuition as a critical component in mathematical thinking, especially in the discovery and development of new ideas. Poincaré believed that intuition was a form of immediate understanding or insight that did not necessarily depend on rational thought processes or logical analysis. He highlights how intuition plays a central role in mathematicians' ability to identify patterns and relationships that are not directly obvious. He argues that although logical and methodical work is necessary to verify and process discoveries, it is often intuition that leads to the most groundbreaking and original insights. This is particularly emphasized in his books "Science and Hypothesis" first published in 1902, which deals with mathematics, space, physics, and nature, and "The Value of Science" first published in 1905, which addresses issues in the philosophy of science.

Poincaré also emphasized the role of intuition in what he called "mathematical invention" or the creative ability in mathematics, where a deep and often unconscious understanding of mathematical objects and their relationships leads to new ideas and theories. He considered this process to be both personal and profoundly creative, distinguishing it from the more strictly logical or computational aspect of mathematics.

Poincaré saw intuition as an essential and powerful aspect of scientific and mathematical work, a source of insight that was both mysterious and fundamentally important for scientific discovery and understanding. Intuition is a guide for discernment and selection of information. From an infinite amount of data, intuition filters out information from various fields of experience and knowledge, organizing information into meaningful patterns on an unconscious level. Intuition is described as an aesthetic sense of beauty and elegance that is verified and studied through logical thinking. (Poincaré 2017)

Albert Einstein

The German-American theoretical physicist Albert Einstein (1879-1955), known for his theories of relativity, valued intuition highly in scientific research. Einstein believed that genuine breakthroughs came from intuitive insights rather than strict logical thinking. Although Einstein emphasized the importance of experimental evidence and rigorous mathematics in science, he argued that significant scientific breakthroughs begin with an intuition. The scientific process involves a balance between intuition and rational thinking, each playing an essential role. Scientific method and rational thinking are of great importance, but there are times when they are insufficient, and intuition can fill this gap. Intuition leads to the most elegant solutions to scientific problems, providing a direct path to the solution with the simplest and most satisfying methodology.

Despite numerous references to intuition, there are hardly any studies on intuition in natural science. There are two qualitative studies, one an indirect empirical study conducted by the Swedish professor emeritus in pedagogy Ference Marton, and the other a direct empirical study conducted by me.

Quantitative and qualitative research use very different methods for collecting, analyzing, and interpreting data. Both have their strengths and weaknesses but are often perceived to be in opposition to each other. Natural science criticizes qualitative research for non-reproducible results and for not defining concepts unambiguously in well-defined categories. Qualitative research has sometimes been used as a preliminary study to formulate hypotheses, which are then explored further with quantitative studies to test and verify hypotheses generated from qualitative research.

The choice between quantitative and qualitative research depends on the purpose, the research questions, and the context in which the research is conducted. In natural science, mainly quantitative methods are used, where numerical data are collected through measurements, experiments or surveys and presented in tables, graphs, or diagrams. The focus is on objective measurements and the use of statistical and mathematical methods for data analysis and conclusions.

The goal is to falsify hypotheses through statistical hypothesis testing, and the results should be generalizable to a larger population. In quantitative studies, quality refers to how well the study measures what it intended to measure and how reliable the study is. The outcome should be the same if the study is repeated. It should be generalizable to other places, other times, and other people.

Qualitative research is primarily used in the humanities and stems from a humanistic scientific tradition with roots in hermeneutics and phenomenology. Qualitative research collects non-numerical data through observations, interviews, diaries, images, or texts. Opinions may be collected through personal interviews and questionnaires from a few individuals or from a case study.

The results are presented as narratives, descriptions or themes. The focus is on exploring and understanding deeper meanings, experiences, behaviors, and contexts. The goal is to describe and interpret various phenomena or social contexts, and the results do not need to be generalized to a larger population. On the contrary, new studies can give rise to new aspects of what is being investigated. In qualitative studies, quality refers to how systematically the study is conducted and how openly it is reported. It should be possible to see how each step in the research is carried out, how the interviews are conducted, and which questions are asked and answered. The choice of concepts and theoretical perspectives should be relevant to the study's goals, and the results should be valuable to those who wish to use them.

39. INDIRECT EMPIRICAL STUDY

The indirect empirical study is based on transcripts of statements about scientific intuition made during group discussions in a program broadcast on Swedish television from 1970 to 1986. Shortly after the Nobel Prizes were awarded, Swedish television used to organize discussions with Nobel laureates in a program called "Science and Man." The program consisted of an hour-long discussion covering topics related to natural science and society, about ten minutes of which were devoted to the question of the significance of scientific intuition. Each time the question was asked, a discussion about the existence and nature of scientific intuition followed. Opinions on scientific intuition were recorded and provided a set of data for empirical investigation on scientific intuition. A total of ninety-three Nobel laureates in medicine, chemistry, and physics participated, of whom three were women with Nobel Prizes in medicine.

For data analysis, a qualitative research method called phenomenography (Marton 1981) was used. This methodology is based on the assumption that there is a third world of "objective content of thought." This world consists of shared perceptions within a given culture, known as the "collective consciousness." It is a designation for the various ideas, perceptions, and beliefs shared within a given culture and society. Between the collective consciousness and the individual consciousness, there is a level where a number of qualitatively different perceptions can be identified for each phenomenon. These perceptions are logically related to each other and define the phenomenon's outcome, the whole of which each perception is a part.

The purpose is to find variations that are common by identifying a limited number of similarities and differences in perceptions. The task is to map different perceptions, compile these various types of statements into description categories, and examine the relationship between these categories in a phenomenographic outcome space. The goal is not to classify and predict but to focus on the qualitative content of the different aspects, not their frequency.

The analysis is based on statements about scientific intuition made by seventy-two researchers who, directly or indirectly, perceived scientific intuition as distinct from logical thinking. Six participants denied experiencing scientific intuition, and ten did not participate or were unsure about the topic. The analysis results showed that the most fundamental aspect of scientific intuition was that researchers chose directions or found solutions to problems for which there was insufficient data in any logical computational sense. The choice of direction was interpreted as reflecting a marginal, not fully conscious, metaphorical perception of the phenomenon as a whole, without knowing its parts. It is a simultaneous grasping of the whole long before knowing its parts in detail.

Scientific intuition is explained as a shift from what cannot be consciously discerned to what is the object of immediate attention. The emergence of an answer is seen as an illumination of what was previously seen in a vague global form. Intuition is characterized by certainty based on a feeling or a perception of an aesthetic or quasi-sensory nature, based on an initial vague global, not fully conscious anticipatory perception of the sought solution. Intuition develops through extended and varied experience of a research object.

Scientific intuition is interpreted as a special case of intuitive understanding. Intuitive understanding is a specific type of intuitive thinking in general, which can be found in very young children when they handle simple arithmetic. It is reminiscent of the Nobel laureates' quasi-sensory perception underlying the intuitive understanding of the phenomenon they handle. The immediate perception of the total problem is called intuitive, in that it is certainty based on the implicit perception of the total problem. Scientific advances are field-dependent, and no one else can make progress except those steeped in a field's information. The paradox is that to expand a field, one must build on it while being able to step outside it. This paradox is resolved by scientific intuition, says Marton (Marton, 1994, p. 469).

40. DIRECT EMPIRICAL STUDY

The direct empirical study is conducted by me. It is based on individual interviews with Nobel laureates about their experience of scientific intuition. The aim is to find variations that are common by identifying a limited number of similarities and differences in perceptions.

The task is to map different perceptions, compile these various types of statements into description categories, and examine the relationship between these categories in a phenomenographic outcome space. The goal is not to classify and predict but to focus on the qualitative content of the different aspects, not their frequency.

For data analysis, the qualitative research method called

phenomenography (Marton 1981) was used. The investigation involves two main steps: individual interviews and analysis of interview transcripts. The sample group consists of researchers who received the Nobel Prize for groundbreaking scientific discoveries in physics, chemistry, and medicine. In the autumn of 1991, I sent a letter to Nobel laureates in natural science who had received the Nobel Prize between 1987 and 1990. The letter included a request for an interview about scientific intuition. About two-thirds of the Nobel laureates responded to my request, indicating they would attend the Nobel Jubilee in Stockholm in 1991. During the Jubilee, I was able to randomly contact ten Nobel laureates who had accepted the interview request.

Interviews

The interviews were conducted in English, recorded, and lasted about half an hour to an hour. The interviews can be described as open and semi-structured. There was an opportunity to delineate and define the content of the interview. The following four questions recurred in all interviews:

- Do you have any experience that you would call scientific intuition?
- How would you describe the experience of scientific intuition?
- What do you think are the conditions for scientific intuition?
- How would you explain scientific intuition?

Analysis

The analysis of the transcribed interviews was carried out in four main steps with different purposes. The first step aimed to get a holistic impression and discover differences in handling the content through an unbiased reading of the interviews. The second step aimed to systematically identify differences and similarities in the statements and contrast the interview responses with each other. By viewing each statement in light of the others, direct quotes about scientific intuition that describe different meanings were identified. The third step aimed to establish different categories that are clearly distinct from each other and not overlapping. The different meanings were grouped into different clusters, called categories. The fourth step aimed to see how the categories relate to each other and if there is a logic in the relationships between the categories. The study is presented in three main categories: outcomes, experience, conditions, and explanations of scientific intuition.

OUTCOME

The outcome of scientific intuition is described both before and during the research process. Scientific intuition results in the perception of something unusual, an initial idea, decision-making, and readiness to think differently.

Perception of Something Unusual

"The intuition part of it is really how to investigate something that doesn't appear to be ordinary. And that is where really advances are made when we look at things that are a little bit peculiar and see if we can explore this a little bit further...What area should I work and explore in, and what area should I not work and explore in. If I see something that looks peculiar or different from what I expect it to, then I would explore it" (Schwartz).

An Initial Idea

"It can be in the beginning of a project, as the beginning idea. Before you start a project you think it out...The intuition part will be in that moment when you decide that is the way I want to do it" (Lehn).

Decision

"Almost every time you decide upon a particular kind of experiment it is a choice and that choice is governed largely by intuition....Whenever you have a choice where logic does not say absolutely what you have to do, you could argue that is intuition. A judgment not based on rigorous logic but more on this accumulated experience" (Lederman).

Readiness to Think Differently

"If you think of something quite different to do, there is where you are relying on intuition. I think it is there it enters strongly. Some way of making your thought different from other people's thoughts up to that time...A willingness to think about a problem in a different fashion" (Ramsey).

Throughout the Research Process

"I think it is all over. In the beginning phases when you have little that guides you and when you are writing up the paper have it all understood and it is released. It can occur at any stage" (Ramsey).

"Intuition comes into every step, when you have started, when you have problems along the way, at the end, the synthesis" (Lehn).

EXPERIENCE

The experience of scientific intuition is described as a feeling of something being right, wrong, useful, insight, and coherence.

Right

"I had the feeling it got to be right. It can't be too far away. But you can't justify it. It is something that appears sort of internally that there is a very good likelihood that you are going to be right...In the early days of electron scattering nobody knew what it would show. In fact, it was said it would not show anything and would only be a waste of time. One of the reasons I was interested in it was the intuitive point of view. I can't say I was convinced that I would see these results...That was such a flaunt point of view at that point of time. If anybody had said that a proton and a neutron had points inside, I said absolutely not. There is not a chance in the world that would be the case. But I knew I had to ask that question because I had sort of an intuitive idea that I had to look at data this way. You do a lot of things with intuition that you can't justify by logic" (Friedman).

Wrong

"If you stand in front of a protein structure it is by no means obvious what the important details of this structure are because the components of the structure may serve very different purposes. To distinguish between these is not trivial at all. It needs a certain feeling that at first cannot be explained in rational terms, it is to identify regions of the structure that are of particular interest...I had from the beginning the feeling that in the arrangement which is the heart of the structure there was something missing in the model that we did. In thinking about it always came back that this cannot be all.

And indeed when we later got more data and more information it became clear that there is indeed an additional element in part of the structure. It is a trivial thing. It was a water molecule that is trapped within the protein. It can provide very crucial interactions within the pigments that are involved in the first steps of the function of the molecule. I would call that in a way intuition to make a reservation about our first model, saying don't regard this as the whole story. It was a sense of having an incomplete picture of some detail not being there that should be there. It also made me reluctant to publish this structure as it is because I thought there is more to it. These feelings have influenced my decisions wherever there is an ambiguous situation" (Deisenhofer).

Useful

"There was a feeling that this was maybe a useful direction to go. That is where the intuition came in, a kind of stubborn feeling that it might be useful. For example, there was a time when we were looking for a particular kind of particle and we got a particular way of looking for that particle and towards the end of that experiment we realized that maybe we could continue this experiment in a slightly different direction. This had no logical basis but it was a correct direction. There was a feeling that this was maybe a useful direction to go. That is where the intuition came in, a stubborn feeling that it might be useful...I would never have believed it was a point-like structure, you see...There is no reason why I should look there but I looked anyway. In science like particle physics, you have to convince a committee. They have to share your intuition. It is very hard to communicate your feeling" (Lederman).

Sudden Insight

"The question that was asked was whether one could investigate the behavior of a certain class of elementary particle interactions in high energies and nobody saw any way of doing it. In that evening I was in bed thinking about it because it seemed to me to be absolutely possible to do it very rapidly. Just instantaneously it was clear what the right way was to do it. It wasn't a matter of making a list of all the alternatives and striking them off a hundred of times. It was a very quick process and it was much like knowing the answers.

So from that point it was just straightforward. This is very different from logical reasoning. This was suddenly one has an idea. What brought the idea I don't know. All I know is that one minute I had no idea and one minute later I knew exactly how to do the experiment. It wasn't a matter of a dreamlike thing. It was a matter of really seeing very quickly that all of the elements that I needed were present in my experience. At that moment I knew exactly how to do it. So it was a matter of finding myself at the end of that thought process without having to go through very much" (Schwartz).

Coherence

"Rather than a feeling, it recognizes that some order has formed, a sudden coherence has come up. Probably this feeling of intuition is the sense of coherence. Something out of the questions you have asked yourself realizes that would be the way" (Lehn).

CONDITIONS

Conditions for scientific intuition are described as crossing experiences and knowledge from different areas, not being bound by knowledge, taking breaks from work, and giving space for initial thoughts.

Cross Experiences and Knowledge from Different Areas

"I think it is very frequently a cross between two experiences that give you the thing...To cross knowledge between different fields" (Ramsey).

"This kind of intuition only comes from different fields. Not from the fields that you are specialized in. Because in the field you are specialized in you normally remember all things and all the details" (Michel).

Not Being Bound by Knowledge

"You do need to have knowledge but you also recognize that you can be caught by it" (Ramsey).

"You need a certain amount of knowledge, not too much. If you have too much knowledge it is a sort of baggage" (Lederman).

"I don't think that knowledge is really very important. I don't have these intuitive feelings very often, but when I have them it may be on subjects that I don't really know much about" (Deisenhofer).

Taking Breaks

"It is less likely to come from what you are just directly working on, a specific problem in your own field. I think you do need to step totally out of the problem for a while for personal relaxation, to get a new perspective" (Ramsey).

"In my case, it seems to be more likely that these things happen when I don't work on that particular problem, because if I work on a particular problem then the pathways of thinking they are following a mainstream of perhaps experimental or theoretical procedures...That is the motivation for me to sometimes interrupt my working on something...because it allows things to come to the surface" (Deisenhofer).

Giving Space for Initial Thoughts

"You can't be too demanding of these initial thoughts. You have to give these thoughts some degrees of freedom, a certain amount of room to move in. If you try to be too logical and too demanding with these things the intuition dies. The initial thought is not necessarily totally right but it is a direction that may lead to the right...Anything you do that tends to liberate your ways of thinking can help in terms of intuition" (Friedman).

EXPLANATION

Scientific intuition is explained in terms of incomplete information, a step in logical reasoning, preliminary conclusions, fuzzy logic, related to creativity and imagination, and the ability to learn.

Incomplete Information

"This information is not complete. There are some relationships in this information and very frequently this information will suggest approaches or points of view that you are not even aware that you have been thinking about. And that is what to some extent is intuition" (Friedman).

Step in Logical Reasoning

"Reasoning is more deductive, but even when you are doing logical reasoning you have to try to think what is the right next step in the logical reasoning. I suppose almost all intuition eventually gets fitted into a logical process" (Ramsey).

Tentative Conclusions

"As I understand the meaning of the word intuition it is the reaching conclusions or tentative conclusions without strict logic" (Lederman).

"Generally, when we talk about logic and reasoning, it means that we can describe exactly what we are thinking at every step. I think that when you make some guesses or you think that you are using your intuition, it is simply that you are not fully aware of how your thought processes are arrived at...or you can't completely justify in words" (Altman).

"It is not logical, conscious reasoning. It may be related to that and it is usually reduced to that" (Cram).

Fuzzy Logic

"One difference between intuition and this strict deductive way of thinking is that intuition is soft. Rather than straight logic, it is fuzzy logic...I am convinced it has the same steps and the same background as reasoning, except that it is a compressed reasoning" (Lehn).

Related to Creativity and Imagination

"To a certain extent, I think that intuition is almost like creativity. Intuition is a way of solving a problem in a different way. They are very much related. Creativity to a large extent is a different side of intuition" (Friedman).

"I think intuition is linked to imagination, the ability to generate many ideas, to represent in your mind different aspects and to be able to go very quickly from one to another and to make bridges" (Lehn).

Ability to Learn

"People have different degrees of intuition. Some people have wonderful, incredible intuition. I do believe that you can teach people to use whatever intuition they have" (Friedman).

SUMMARY AND DISCUSSION

In summary, scientific intuition refers to events both before the decision of a new research project and during all stages of the research process until a solution is achieved and everything is integrated and understood. Scientific intuition manifests as the perception of something unusual, an initial idea, decision-making, and a readiness to think differently. It is experienced as a feeling that something is right, wrong, useful, insightful, or coherent. Researchers choose to follow a particular research path based on a feeling, even if there is no concrete evidence supporting this path. It is a sense that there is a good chance it may turn out to be the right path. Sometimes, researchers can find answers to their questions or solutions to problems through a sudden insight without having to try different alternatives. It is a very rapid recognition of all the necessary elements. Suddenly, there is an answer, but how that answer came about is unknown.

Conditions for scientific intuition include crossing knowledge and experiences from different areas, not being bound by knowledge, taking breaks from work, giving space for initial thoughts, and the ability to learn. It is important to stop thinking about the problem to gain a new perspective. Breaks from work are needed to allow new ideas and initial thoughts to emerge. There must be room for ideas and thoughts to move freely and fertilize each other. Being too logically demanding leads to a loss of intuition.

Scientific intuition does not originate within one's field of expertise. Working on problems within one's field means that thought pathways follow established experimental and theoretical procedures. It is when researchers are not working on a specific problem within their field that there is a likelihood of scientific intuition. Too much knowledge is a burden because intuitive feelings often arise in subjects the researcher is not familiar with.

Scientific intuition is explained in terms of incomplete information, a step in logical thinking, preliminary conclusions, fuzzy logic, related to creativity, imagination, and the ability to learn. The information is not complete, but it can suggest perspectives that the researcher has not considered before. Intuition is not the same as logical thinking but a step in logical thinking. If logical thinking generally means that we can describe exactly what we are thinking at each step, intuition is a decision for the next step in logical thinking. Logical thinking is deductive, but even when we think logically, we must decide what the next step in logical thinking is. Scientific intuition complements logical thinking and does not exclude it. If scientific intuition leads to new ideas and insights, logical thinking and scientific methods are used for verification.

The results indicate that intuition plays a key role in scientific discoveries. Intuition is a way to approach problems from new perspectives. It begins with the perception of something unusual, which is meaningful to explore further. This can involve exploring anomalies in a widely accepted theory or investigating something that seems interesting and relevant. It is a sense of meaning that drives the researcher's decision to initiate a new research project. A sense of meaning is a strong driving force and worth supporting. If researchers perceive something that feels meaningful to explore further, but which falls outside the scientific paradigm, they must invest resources and time themselves.

The natural scientific paradigm is based on a rational empirical knowledge paradigm. Knowledge should be secured through rational thinking and empirical measurements. It is no longer sufficient to secure knowledge. Primarily, is to find out what are meaningful objective measurements, not just to make objective measurements. It is humans who choose what is meaningful. The amount of available information is so extensive that we do not know what is important to pay attention to. We are overwhelmed with so much information that it is impossible to sort and evaluate what is fact and what is not. We need to be able to sift through all the information, not acquire as much fact and information as possible. We need to determine what is meaningful information and discern what is reliable and unreliable information.

The growth of artificial intelligence means that many of human intellectual functions are performed by intelligent machines. This opens up the possibility to focus on human experiences of meaning, creativity, and imagination. It is through human experiences of meaning, creativity, and imagination, that new models for understanding the world are created and technical innovations are made. We can expand and extend our awareness in areas that we previously thought were pure fantasy. We live in a universe that creates billions of life forms with infinite creative imagination. Humans are also a manifestation of the universe. The universe creates through humans, and we become its co-creators by creating art, music, literature, and technical inventions.

Creativity often refers to the ability to generate new ideas, but less frequently to the value and meaning of these ideas. In the measurement of creativity, a distinction is made between creativity and intelligence. Creativity tests measure the ability to think innovatively, while intelligence tests measure the ability to think logically. Recent research indicates that there is no clear distinction between creativity and intelligence in the natural sciences. Creative problem-solving and production involve an interplay between intuitive and logical thought processes.

Metaphorically, the relationship between intuition and logical thinking can be described in terms of the Chinese symbols Yin and Yang. In Chinese philosophy, Yin and Yang are two polarities that are interconnected and form a whole. They are mutually dependent on each other and create conditions for each other. Yin represents a soft, open, and passive principle. Yang represents a firm, closed, and active principle. Yin always contains a part of Yang, and Yang always contains a part of Yin. Similarly, intuition and logical thinking are two polarities that are interconnected and form a whole. They create conditions for each other and are mutually dependent. Intuition always contains a part of logical thinking, and logical thinking always contains a part of intuition. They can be adapted to each other so that they interact in a balanced way.

Intelligence is not only about logical thinking but also about understanding. Without understanding, intelligence is merely mechanical computational processes. Intelligent machines are good at calculations and instruction, but humans provide understanding and meaning. Konceptual and intuitive understanding are two different ways of understanding are conceptual. When we understand something conceptually, we draw conclusions and recognize patterns through active thinking. It is a slow and conscious thought process, where we rely on logical thinking for judgment and decision-making. It provides objective knowledge and judgment, which comes through reflection and analysis of a subject or idea. Conceptual understanding involves knowing something by relating something conceptually to another conceptually and is based on sensory or perceptual content. We can understand something theoretically, but it says nothing about skill. For example, we can learn a theory to get a driver's license, but it does not provide the skill to drive a car. Theoretical understanding says nothing about how it feels. Knowing how honey tastes is entirely different from having theoretical knowledge of what honey is.

When we understand something intuitively, we draw conclusions and recognize patterns without active thinking. It is a quick and unconscious thought process where we rely on feelings for judgment and decisions. Intuition is subjective knowledge and judgment and does not require reflection and analysis. It is a way of knowing something through feeling, without being able to put it into words or knowing where this knowledge comes from. Intuition provides emotional understanding and is not based on sensory or perceptual content. Intuition is part of the human psyche and a key component in how we gather information, understand the world, and make decisions. Intuition breaks established patterns of thinking, initiates research, explores new thoughts and ideas, and provides new perspectives on problem-solving. Intuition is particularly useful for judgment and decision-making in situations with complex or insufficient information. Everyone has the capacity for intuition and can develop it.

A question arises whether human intelligence has a value that cannot be replaced by an intelligent machine. Proponents of artificial intelligence argue that intelligent machines are superior to humans in rational thinking and decision-making. By constructing algorithms based on formal logic, calculations and decisions can be made much more efficiently and safely than by a human. We are not governed by formal rules for how calculations should be made and decisions taken. We are often more loyal to our feelings and beliefs than to rational analysis.

Artificial General Intelligence (AGI) is an agent that is said to possess intelligence that far surpasses human intellectual capability in virtually all areas. AGI has the ability for deep learning, based on how our brain learns through practice and feedback. This leads to self-learning algorithms that create super-algorithms, with their own goals and decision-making.

Some researchers in artificial intelligence claim that it is possible to create conscious AI systems. They are conscious in the sense that they can perceive the environment, take action, and maximize chances of achieving goals. A robot that can interpret signals from the environment is considered conscious. Others advocate the possibility that future AI systems can be conscious, even if they are not sensory bodies or robots, but only exist as future software.

A body is needed to receive signals from the environment and make electromagnetic radiation into a recognizable pattern. Our bodily activity, our ability to move, breathe, and speak, is essential for interaction with environment. The body is necessary to perceive the environment; otherwise, information is only about what happens in a system and is like a dream state. When we are in a dream state, we are unaware of the external world. In the dream, it is only about information in the brain, and no interpretation of signals from the environment.

It is impossible to determine if a robot is conscious based on behavior. Behavior does not say anything about what it feels like to be conscious. If something behaves as if it is conscious, it is not necessarily an indication that it is conscious. Robots can perform activities that we designate as conscious, but we cannot know if a data-driven robot is truly conscious just because it behaves as if it is. There are still no objective criteria for knowing if robots have experiences that can be described as conscious. Consciousness, as described by AI, does not explain why there are subjective experiences and a sense of self. The question of whether robots can have subjective experiences and self-awareness remains unanswered.

Consciousness is so intimate that we do not notice it. We do not think of consciousness as something significant, yet we live and move in consciousness just as fish live and move in water. We do not need to seek consciousness because it already exists in humans. Consciousness is so natural that even natural science has ignored its significance. Natural science provides knowledge about the physical world through our ability to think, but we cannot imagine or explain consciousness based on any conceptual framework, for consciousness exists before we start thinking and conceptualizing. Consciousness is often overshadowed by thinking. It is invisible to the eye but a prerequisite for all sensory impressions, thoughts, and feelings. All experiences happen in consciousness. Without consciousness, there are no subjective experiences of color, shape, warmth, cold, pain, taste, movement, smells, or sounds.

Our intellect is limited, bound to forms and time. When we use the intellect, we can only think conditionally based on experience and knowledge. The only place that is free from thinking conditionally based on experience and knowledge is consciousness. Consciousness is unlimited, formless, and timeless. The intellect is good at sorting, storing, and preserving information but not at generating new ideas. It is developed for survival and seeks patterns and routines. The intellect can provide myriad theories from different perspectives, but never a holistic perspective. Only consciousness can give an overarching perspective and make us aware of something new. We tune into a creative intelligence with infinitely more possibilities than the intellect by breaking the intellect's limited way of thinking,. It is an intelligence that goes beyond but encompasses the intellect and gives rise to all innovations in science, art, and technology.

We identify with our thoughts and believe that thoughts are the only reality we can be certain of, but thoughts have no inherent reality. Thoughts are fleeting and changeable, yet we give them great power. Consciousness is the only constant we can know for sure. When we break our identification with thinking, we discover consciousness, which lies deeper than our thoughts. Consciousness is always present but obscured by thinking. When we understand that our identity and sense of self come from consciousness ans not from our thoughts, our way of relating to the world changes. This transformative shift offers a new perspective on ourselves and the world. We can understand that our deepest identity is in consciousness through introspection and logical analysis, exploring what happens when we do not think. To gain direct experience, we need only to quiet all thoughts. What remains when thoughts are quiet is pure being.

Thinking has an important place, but when it occupies all our attention, it creates an imbalance. Our entire world then stems from thoughts, leaving no room for simply being. Thinking absorbs everything in our consciousness and transforms it into thoughts. There is no problem with thinking, but not being aware of the present is a problem. How we attend to the present determines the quality of our experiences.

When we do things, we need to be more present, not just perform tasks mechanically. This brings more quality to our experiences. When we are fully focused on what is happening in the present, we are in touch with being. This is being in complete mental presence and awareness, known as conscious presence. Conscious presence is characterized by being alert combined with calmness that arises as thoughts quiet down. It is a heightened state of wakefulness, different from ordinary wakefulness, and is the highest form of wakefulness, implying vigilance and readiness for anything. .

Just as we cannot force ourselves to sleep, we cannot force conscious presence. The inability to stop thinking can be a sign that thinking has become compulsive, creating a dependency on thinking that can turn into an obsession with constant thought. However, we can rise above our thoughts and observe them. When we observe our thoughts, we are no longer identified with them, and thinking loses its hypnotic power. Thinking spontaneously ceases, by shifting attention from thinking to conscious presence, This shift releases energy and makes us more aware. The body fills with energy and feels more alive. In conscious presence, we can enjoy our existence without any dependency. We can be active, explore, and learn new things without a sense of necessity or compulsion.

There is a constant background noise of inner self-dialogues within us, often unconscious. This noise consists of involuntary, repetitive, and irrelevant thoughts on the verge of unconsciousness. They distract us, create unnecessary stress, tire us out, and steal vital energy. These self-related thoughts easily trap us in thought loops, making them hard to stop and preventing us from relaxing. When we are free from self-dialogues, there is only pure being, a sense of life, and aliveness. We can learn to do things without losing ourselves in thinking and doing. When we lose ourselves in thinking and doing, we can always return to being.

Many people are only aware of doing and unaware of being, which creates stress. Stress arises when we want to do as much as possible in the shortest possible time or do many things simultaneously. We can do things more slowly, taking one step at a time, recognizing that each step is important in achieving goals. We work more efficiently when not filled with thoughts about the goal. Life is not just about thinking and doing but also about being. We need to slow down and allow ourselves to simply be. We often give only partial attention to what we do. Full attention requires training in just being.

We can explore and understand the human psyche at the brain level, but we do not know what consciousness is. We can see in the brain if we are conscious or not, but that does not explain what consciousness is. We cannot explain consciousness conceptually because all concepts presuppose consciousness. We can only experience it by quieting all thoughts and giving up our attempts to understand it conceptually. Consciousness cannot be explained conceptually, but we can describe different states of consciousness.

We can never understand reality as it is through thinking because our brain interprets information based on knowledge, experiences, and expectations. We believe we are separate from an external world and can observe the world independently of ourselves. We perceive ourselves as separate from the universe, forgetting that we are also a manifestation of the universe. It is an illusion to think we can stand outside and observe an external world separate from us. The world we see is not independent of our mind. We see a world filtered by our senses and conditioned by our experiences and thinking. Only consciousness is free from the conditioning of experiences and thinking.

We identify with our mind and live as in a dream state. In a dream, we experience ourselves as a person around whom things happen, convinced that everything happening is real. Every scene in the dream manifests without cause, and nothing happens in reality. When we wake up from the dream world, we realize that everything is just a projection of consciousness onto itself. Even in the waking state, we experience ourselves as a person around whom things happen and are convinced that everything is real. Similarly, we can wake up from the waking state and realize that everything is just a projection of consciousness onto itself. If the mind can create something that feels real in the dream state, perhaps the mind can also create something that feels equally real in the waking state.

The world we see is shaped by our organic senses and the thoughts we use to interpret it. We do not perceive the world as it is but as useful for promoting our survival. Our perceptions act as a filter for the world. We understand through the limitations of our senses and evaluate, judge, and perceive the world conditionally.

Our life is the dominant story in our mind. It is through this story that we interpret everything we encounter: situations, events, and other people. We construct and organize experiences into a coherent picture of ourselves and the world. From an early age, we begin to develop personal perceptions or thought structures that shape behavior. These are representations that form the basis for how we interpret, think, remember, and make decisions. We think, perceive, and decide through structures conditioned by our personal history and cultural identity. Because these are tied to our personal identity, they have an element of permanence, making them difficult to change.

Our observations are based on past experiences and knowledge, influenced by personal perceptions, and become what we automatically observe and react to. We think in patterns consistent with our expectations, preconceptions, rooted in experience and knowledge. This leads to an over-generalized way of thinking, acting as a filter for incoming information. These patterns of thinking and perception lie on the edge of the unconscious. We rarely notice them, and they are not controlled by will. Even if we do not notice them, they provide basis for unconscious decisions. Unconscious thoughts govern our lives, and we act on them.

We often start from the mistaken assumption that more thinking will lead to a new idea or thought. On the contrary, it is when we break the stream of thoughts and stop actively thinking that the possibility opens up to perceive a new idea or thought. For a new idea or thought to emerge, space is needed. We can create a gap in the stream of thoughts. In the brief interval between thoughts, we are momentarily conscious without thoughts. The conscious field between thoughts is a space free from thinking based on experiences and knowledge. There are many more opportunities for a new idea or thought to appear when consciousness is not filled with thoughts.

New ideas and thoughts cannot be thought out or calculated. Thinking is limited because it is always based on some unspoken assumption we take for granted. It is not thinking but consciousness that is the source of new ideas and thoughts. Consciousness is not based on any assumption other than itself. By questioning established assumptions and exploring new ones, great creative breakthroughs occur. The starting point for knowing something always begins with not knowing but wanting to know something. It is when we acknowledge that we do not know and ask a question about what we wish to know that there is potential for new ideas and thoughts to arise spontaneously and naturally. By stopping thinking and being comfortable with not knowing, new ideas can emerge in consciousness.

All creation and all creative breakthroughs in science, technology, and art come from non-thinking and inner calm. When we want to get a new idea or thought, we can let go of thinking and just be aware. We can take a short break so that one thought does not immediately lead to a stream of thoughts. By taking a short pause, we can gain more energy to use our thinking more focusedly when seeking a creative solution.

Scientific knowledge development is increasing exponentially and tends towards greater fragmentation and specialization. Each research area develops its methods and highly specialized languages, making it difficult for researchers to communicate across scientific boundaries and obtain the overview necessary for groundbreaking creative breakthroughs. The availability of knowledge is so vast and changes so rapidly that it is impossible to keep up with the forefront questions within one's own research area, let alone monitor questions in other fields.

Our traditional paradigms for knowledge development and education are no longer adequate. Society changes so quickly that it is impossible to know what knowledge and skills will be required in the future. The amount of information is so extensive that we do not know what is important to pay attention to. We are overwhelmed with so much information that it is impossible to sort and evaluate what is fact and what is not. The goal is no longer to acquire as much information and facts as possible, but to sift through the information and determine what is meaningful. We need to discern what is reliable and unreliable information, not acquire more information. The idea that knowledge is power is outdated because information is unlimited. Education focused on acquiring all knowledge within a specific field is no longer sustainable.

Generative AI has changed our perspective on education and can transform it in many ways. This technology creates a more personalized learning environment, allowing everyone to reach their full potential and stimulating creative problem-solving and new skills, regardless of background and abilities. However, artificial intelligence can never replace the creative intelligence found in humans, though AI can be very useful by doing what it excels at and what we cannot.

As our dependence on artificial intelligence increases, self-awareness becomes important. Physics has given us great technological inventions, but without understanding ourselves, technology can take authority over us. We are the ones who understand what is right and wrong and can create a better world using the creative intelligence and free will present in consciousness.

All life forms evolve by facing challenges and overcoming obstacles. There are two main types of obstacles: those in the outer physical world, involving the development of our intellectual and physical skills, and those in the inner mental world, involving liberation from the ego's involvement in the intellect. The intellect is a fantastic instrument, but when we let the ego infiltrate our intellect, thinking becomes dysfunctional. This leads to a dysfunctional world and ultimately to the destruction of the human world. The ego expends more energy attacking and justifying beliefs than seeking creative solutions to problems. It uses illogical explanations, excuses, and denial of real problems. These strategies affect our way of thinking and making decisions, shifting focus from real problems to pseudo-problems.

The development of AI-based language models raises the question of the value of knowledge. The value of knowledge is that it should lead to changes in behavior and society in some respect. Historical knowledge that does not lead to any change becomes merely a container for what we already know and quickly loses its relevance. With the accelerating growth of knowledge, we are less able to make predictions for the future and cannot use the present in a more meaningful way. Change and development occur in the present, not in the past or future. It is in the present that we get new ideas and insights.

The French philosopher and mathematician René Descartes (1596-1650) is known for his famous statement: "je pense, donc je suis" (I think, therefore I am). In this statement, he saw certain knowledge that could not be doubted. He concluded that all necessary and immutable truths must belong to humans independently of sensory experience. Our senses deceive us and cannot provide us with complete certainty. Our sensory knowledge is always temporary and changeable. The concept of intelligence has evolved to also include creativity and consciousness. Another more appropriate version might be to say: "I am aware, therefore I can think, feel, and imagine."

Through Descartes, rational thinking became a source of knowledge of a higher nature than what our senses provide. Rational thinking is still considered by many to be the highest cognitive function a human can achieve. The great breakthrough in natural science during the Enlightenment was rational thinking. It is not the highest potential that a human can reach. The next great breakthrough is to discover and unleash the creative intelligence present in all humans.

Artificial intelligence can never replace but can enhance human creative intelligence. It can process and analyze large amounts of data quickly and efficiently, identifying patterns and relationships that can lead to new discoveries. It can suggest hypotheses, provide overviews of different areas, explore the consequences of various choices, and help researchers stay updated on the latest developments in different research fields. It can come up with new ideas and solutions to problems within the framework of programming and training, but it is not as dynamic and intuitive as human creative intelligence. Humans can think outside the box and create something that has not existed before. Although artificial intelligence can demonstrate aspects of creative intelligence, it does so based on predetermined algorithms and pattern recognition, not through awareness and intuition. It is a powerful tool for human creative intelligence but cannot make groundbreaking scientific discoveries on its own.

The exponential development of artificial intelligence that characterizes our contemporary era forces us to reassess what truly constitutes consciousness. The development of new technology and an increased understanding of consciousness have raised questions about the consequences of technical and societal changes. Each step of new technology impacts our way of living. First came the art of writing, which gave us the ability to record all our thoughts. Then came the printing press, which allowed us to promote the ideas of individual freedom, equality, and justice. Then came Descartes, who argued that it is through rational thinking that we gain all the knowledge we need. Then came the discovery of the computer and the internet, which provided access to what everyone else is thinking. Now comes the discovery of creative intelligence, which is the foundation for all development and new groundbreaking discoveries

REFERENCES

Alexandersson M. (1994). *Den fenomenografiska forskningsansatsens fokus.* I Starrin, B. & Svensson, P. *Kvalitativ metod och vetenskapsteori.* Lund: Studentlitteratur.

Amabile, T. M. (1982). *Social psychology of creativity: a consensual assessment technique.* Journal of Personality and Social Psychology, 43(5) p.997-1013.

Amabile, T. M. (1983) *The social psychology of creativity.* New York: Springer-Verlag.

Bateman, A., & Holmes, J. (1995). *Introduction to psychoanalysys: Contemporary theory and practice.* New York: Routledge.Bauby J.D. (2008). *Fjärilen i glaskupan.* Albert Bonniers förlag.

Benson H. (1975, 2001). *The relaxation response.* Harper Collins. USA.

Bergson H. (1992). *Introduktion till metafysiken.* Förlag Pontus.

Bergson H. (2002). *The Creativ Mind.* Dover Publications Inc.

Biehler R., Snowman J. (1997). *Psychology Applied to Teaching.* Publisher: Houghton Miffin Harcourt(HMH); 8th edition.

Binet A. (1890). *Percepttions dénfants och recherches sur les mouvement de quelques jeunes enfents.* La Revie Philosophique.

Bohm D. (1980). *Wholeness and the Implicate Order.* London Routledge.

Brun J. (1994). *Platon och Akademin.* Alhambra Förlag AB.Bräutigam (Heinstedt) E. (1972). *Effekten av transcendental meditation på narkotika missbrukare. Psykologexamensarbete.* Psykologiska institutionen. Lunds universitet.

Bräutigam (Heinstedt) E. (1973). *The effects of transcendental meditation on drog abuse.* Paper presented at International Congress for Applied Psychology: Symposium on Transformations of Consciousness, Montreal Canada.

Bräutigam (Heinstedt) E. (1975). *The Physiological and Psychological effects of Meditative States of Relaxation.* Paper presented at the 3rd Congress of the International College of Psychosomatic Medicine, Rom Italy.

Bräutigam (Heinstedt) E., Bräutigam A. (1975). *Skewness Changes in the EEG during Mental Divergent and Convergent Produktion.* Psychologial Research Bullentin. 15 :11.

Burkeman O. (2021). *Four thousend Weeks. Time and How to Use It*. Peguin Random House UK.
Buss D. (2017). *Evolutionary Psychology:The new science of the mind.* Psychology Press Ltd.

Carrol S. (2016). *The Big Picture: On the Origins of Life, Meaning, and the Universe Itself.* Förlag: Dutton.

Carrol S. (2019). *Something Deeply Hidden. Quantum Worlds and the Emergence of Spacetime.* Förlag: Penguin Publishing Group.

Chalmers D. (2021). *Philosophy of Mind.* Oxford University Press Inc.

Chomsky N. (1975). *The Logical Structures of Linguistic Theory.* Kluwer Academic/ Plenum Publishers

Chomsky N. (1957). *Syntactic Structures*. Martino Fine Books.

Csikszentmihalyi M.(1996). *Creativity. Flow and the psychology of discovery and invention.* New york: Harper Perennial.

Csikszentmihalyi M. (2000). *Beyond boredom and anxiety: Experiencing Flow in Work and Play.* Förlag Jossey-Bass Inc.,U.S.

Custance J. (1951). *Wisdom, Madness and Folly:the philosophy of a lunatic.* London: Gollancz.

Damasio A. (1994). Descartes' Error: Emotion, Reason, and the Human Brain. Penguin Books.

Damasio A. (2021). Feeling and Knowing. Making Mind Conscious. Little, Brown Book Group.

Danielsson U. (2020). *Världen själv.* Förlag: Fri tanke, Stockholm.

Danielsson U. (2020). *Tidens Ände. Förlag:* Fri tanke, Stockholm.

Darwin C. (2009). *Om arternas uppkomst genom naturligt urval.* Natur Kultur Allmänlitteratur.

Davanger S., Eifring H. Hersoug A. (2008). *Fighting Stress -Rewiews of meditation research.* Acem Publishing, Oslo, Norway.

Davidson D. (1963). *Actions,reasons and causes.* Journal of Philosophy.

Davidson, R. J., & Kaszniak, A. W. (2015). *Conceptual and methodological issues in research on mindfulness and meditation.*American Psychologist, 70(7), 581–592

Dawkins R. (1976). *Den själviska genen.* Oxford University Press.

Dawkins R. (1982). *The Extended Phenotype.* Oxford University Press
De Bono E. (2016). *Six Thinking Hats.* Förlag: Peguin Life

De Bono E. (1990). *Lateral Thinking.* Förlag: HarperPerennial.

Davies P. (1987). *The Cosmic Blueprint.* Förlag: Simon and Schuster

Davies P. (1992). *The mind of God*, Simon and Schuster.

Davies P. (2007). *The Goldilocks Enigma: Why Is the Universe Just Right for Life,* Houghton Miffin Harcourt.

Descartes R. (2005). *Discourse on Method and the Meditations.* Förlag: Peguin Classics

Edling C. (2003). *Kvantitativa metoder: grundläggande analysmetoder för samhälls-och beteendevetare.* Studentlitteratur.

Evans D. Zarate O. (2010). *Evolutionary psychology.* Icon Books Ltd

Faggin F. (2021). *Silicon: From the Invention of the Microprocessor to the New Science of Consciosness.* Förlag: Waterside Productions.

Feldman L. (2018). *How Emotions Are Made: The Secret Life of the Brain.* Pan Books förlag.
Feldman L. (2021). *Seven and a Half Lessons About the Brain.* Picador.

Finke, R. A., Ward, T. B. & Smith, S. M. (1992). *Creative cognition: Theory, research, and application.* MIT Press, Cambridge: MA.

Fishbein H. (1987). *Intuition in Science and Mathematics.* Kluwer Academic Publishers.

Frege G. (2002). *Aritmetikens grundvalar.* Bokförlaget Thales.

Freud S. (2005). *Freud´s Requiem.* Publisher Riverhead Hardcover.

Freud S. (2008). *Freud. Psykoanalysens teknik.* Förlag: Natur Kultur Akademisk.

Frostin A. (2000). *Om tanke:om kognitiv terapi och kognitivt förhållningssätt.* Lundbeckbiblioteket.

Gardner H. (1983). *Frames of Mind - The Theory of Multiple Intelligences.* Förlag: Basic Books

Giulio T.(2012). *Phi. A voyage from the brain to the soul.* Förlag: Pantheon.

Giulio T. (2018). *Sizing Up Consciousness: Towards an Objectiv Measure of the Capacity for Experience.* Oxford University Press.

Goff P. (2019). *Gallileo´s Error. Foundations for a new science of consciousness.* Förlag: Rider Co.

Goleman D., Davidson R. (2017). *The Science of Meditation. How to Change Your Brain, Mind and Body.* Förlag: Penguin Life.

Guilford J.P. (1967). *Nature of Human Intelligence.* McGraw-Hill Inc, US

Goldstein R. (2005). *Ofullständighet: Kurt Gödels bevis och paradox*. Nora:Nya Doxa.

Graziano A. & Raulin M. (2013). *Research Methods: A Process of Inquiry*. Förlag: Pearson Education Limited.

Guilford J. P. (1968) *Intelligens, Creativity and Their Educational Implications.* Publisher: Robert R. Knapp.

Gustavssson B (2000). *Kunskapsfilosofi. Tre kunskapsformer i historisk belysning.* Förlag Wahlström & Widstrand.

Gustavssson B (2002). *Vad är kunskap. En diskussion om praktisk och teoretisk kunskap.* Förlag: Myndigheten för skolutveckling.

Gädernfors P. (2010). *Intuition är implicit kunskap.* Gästinlägg. Micco.se

Gärdenfors P. (2022). *Hur orden får mening*. Natur Kultur Allmänlitteratur.

Hauser, M. D., Chomsky, N., & Fitch, W. T. (2002). The faculty of language: what is it, who has it, and how did it evolve?. science, 298(5598), 1569-1579.

Harari Y. (2018).*Sapiens: en kort historik över mänskligheten.* Förlag: Natur Kultur Allmänlitteratur.

Harari Y. (2022). *Homo Deus: A Brief History of Tomorrow.* Förlag: Natur Kultur Allmänlitteratur.

Heidegger M.(1984). *The Metaphysical Foundations of Logic.* Förlag: Indiana University Press.

Heidegger M.(2019). *Vara och Tid.* Bokförlaget Daidalos.

Heinstedt E. (1990). *Divergent Production – An ”Intuitiv” Cognitiv Style.* Licentiatavhandling. Psykologiska institutionen. Lunds universitet.

Hof W. (2021). *Wim Hof – Andningen kylan och livet.* Volante.

Hoffman D. (2019). *How evolution hid the truth from our eyes*. Peguin Books.

Hume D. (1978). *A Treatise of Human Nature.* Förlag: OUP Oxford.

Hume D. (2004). *Om passionerna – Avhandling om den mänskliga naturen*. Bokförlaget Thales.

Husserl E. (2001). *The Shorter Logical Investigations.* London and New York: Routledge.

Husserl E. (1989). *Ideas pertaining to a Pure Phenomenology and to a Phenomenological Philosophy.* Second Book. Trans. Richard Rojcewicz and André Schuwer. Dordrecht and Boston: Kluwer Academic Publishers.

Huxley A. & Osmond H. (2018). *Psychedelic Prophets: The letters of Aldolf Huxley and Humphry Osmond.* McGill-Queen´s University Press.

Jung C.G. (1932). *The Psychology of Kundalini Yoga.* Notes of a seminar by C.G. Jung with S Shamdasani. 1996 ed. Princeton, N.J. Princeton University Press.

Jung C. G. (1990). *Arketyperna och det kollektiva omedvetna.* Vol.9, Del I av de samlade verken, Princeton University Press, 1990, s. 123.

Jung C. (1992). *Jung to live by.* Published by Warner Books Inc. New York.

Jung C. (1993). *Jung i våra Liv.* Praktisk tillämpning av CG Jungs teorier. Svenska Dagbladets förlag AB.

Jung C.G. (1998). ”*Synkronicitet*”. Bloombury Publishing PLC

Kahneman D. (2000). *Choices, Vales, and Frames.* Förlag: Cambridge University Press.

Kahneman, D. (2017). *Thinking fast and slow.* Förlag: Volante

Kant I. (2004). *Kritik av det rena förnuftet.* Stockholm: Thales.

Kastrup B. (2014). *Why Materialism Is Baloney.* Förlag: John Hunt Publishing.

Kastrup B. (2019). *Idea of the World.* Förlag: Iff Books.

Kastrup B.(2021). *Decoding Schopenhauers Metaphysics.* Förlag: John Hunt Publishing.

Kaufman G. (1980). *Problemlösning och kreativitet.* Cappelens forlag Oslo.

Klein G .(2012). *Nya tankar om kreativitet och flow.* Norstedts, Stockholm.

Klein S. (2008). *Lyckoformeln: Vad vetenskapen kan lära oss om lycka.* Förlag: Natur & Kultur Akademisk

Kuhn T.S. (1962, 2012).*The Structure of Scientific Revolutions.* Continuum University of Chicago Press
Kurzweil R. (2000). *The Age of Spiritual Machines.* Förlag: Penguin Books

Kurzweil R. (2006). *The Singularity Is Near: When Humans Transcend Biology.* Förlag: Duckworth.

Lance W. Will R. (2021). *Evolutionary psychology.* Cambridge University Press

Larsson H. (1892). *Intuition: några ord om diktning och vetenskap.* Santérus Förlag.

Libet B. (1993). *Neurophysiology of Consciousness:Selected Papers and New Esseys.* Biekhauser Boston Hardcover.

Libet B. (2005). *Mind Time:The temporal factor in consciousness.* Harward University Press.

Marton, F. & Booth, S. (2000). *Om lärande.* Lund: Studentlitteratur.

Marton, F. Dahlgren, L-O. Svensson, L. & Säljö, R. (1977, 2009) *Inlärning och omvärldsuppfattning.* Norstedts, Stockholm.

Mednick S. (1962). *The associative basis of creativity.* Psychological Review. 69 (3):220-232.

Mele A. (1997). *The Philosophy of Action.* Oxford University Press.

Mill J.S. (2012). *A System of Logic.* Benediction Classics.

Muller R. (2016). *The Physics of Time.* W.W. Norton.

Nagel T. (2012). *Mind and Cosmos.* Oxford University Press

Osborn A. (1942). *"How to Think Up",* New York, London: McGraw-Hill Book

Osborn A. (1948). *Your Creative Power.* New York, London: Scribner's Sons

Patton P. (2021). *One World, Many Minds.* Scientific American.

Penrose R. (1995). *Shadows of the mind.* Random Hous UK.

Piaget, J. (1970). *Genetic epistemology.* Columbia University Press.

Planck M. & Niels Bohr (2019). *Qantum Theory.* Förlag; Flame Tree Publishing.

Platon (1986). *The Dialogues of Plato*. Förlag: Bantam USA.

Platon (2017). *Platon Scrifter. Bok 3, Staten.* Bokförlaget Atlantis.

Podvoll E. (1991). *Galenskapens förförelse: Om psykosens väsen.* Natur & Kultur. Stockholm.

Poincaré H. (2017). *Science and Hypothesis.* Bloomsbury Publishing PLC.

Popper K.1997, 2016). *Objective Knowledge an Evolutionary Approach*. Stanford Encyclopedia of Philosofy.

Popper K. (2002). *"The Logic of Scientific Discovery".* Routledge Classics.

Popper K. (2017). *Det öppna samhället och dess fiender.* Förlag: H:ström Text Kultur AB.

Popper K. (2020). *Det öppna universum: ett argument för indeterminism.*
Förlag: Vulkan.

Rothenberg A. (1996). *"The Process of Janusian Thinking in Scientific Creativity"*. Creativity Research Journal. Vol 9, Nos. 2& 3. 207-231.

Rovelli C. (2021). *Helgoland:The Strange and Beautiful Story of Quantum Physics*. Peguin Books Ltd.

Rovelli C. (2022). *Helgoland: Den relationella tolkningen av kvantfysiken.* Fri Tanke.

Sartre J.P. (2003). *Léxistentialisme est un humanisme.* Utgivare: Editions Gallimard.

Sartre J.P. (2014). *Das Sein und Das Nichts.* Utgivare: Walter de Gruyter.

Sacks O. (1998). *The man who mistook his wife for a hat.* Brombergs förlag

Sebell D. (2021). *Xenophon´s Socratic Education.* University of Pennsylvania Press.

Seth A. (2021). *Being You. A New Science of Consciousness*. Faber förlag.

Schopenhauer A. (2000). *The World as Will and Representation, Vol.1.* Dover Publications Inc.

Schopenhauer A. (2020). *Världen som Vilja och föreställning. Vol.1.* Förlag: h:ström-Text Kultur AB

Sternberg R. (1985, 2003). *Beyond IQ. A triarchic Theory of Successful Intelligence.* New York City: Cambridge University Press.

Sternberg, R. J. (1988). *Mental self-government: A theory of intellectual styles and their development.* Human Development. 31 (4): 197–224.

Sturemark C. Hofstadter D. (2021). *Konsten att tänka klart.* Förlag: Fri tanke.

Taylor C. (2010).*The atomists, Leucippus and Democritus.* University of Toronto Press.

Taylor J. (2016). *Min stroke.* Förlag: Bra Böcker.

Taylor J. (2009). *My Stroke of Insight: A Brain Scientist´s Personal Journey.* Peguin Books

Tegemark M. (2018). *Life 3.0. Being Human in the Age of Artificiell Intelligence.* Brockman Inc. Svensk utgåva:Volante.

Tolle E.(2017). *Lev livet fullt ut.* Massolit Förlagsgrupp AB.

Tolle E. (2015). *En ny jord: Ditt inre syfte*. Förlag: Massolit.

Tolle E. (2022). *A new Earth: Create a Better Life.* Penguin Books Ltd.

Unwerth M. (2005). *Freud´s Requiem.* Förlag: Riverhead Mardcover

Vygotsky L. S. (2001). *Tänkande och språk.* Bokförlaget Daidalos.

Wallace R.K.(1970). *The physiological effects of Transcendental Meditation: A proposed Fourth Major State of Consciousness.* Ph.D. Theses. University of California. Los Angeles.

Wallace R.K. at al. (1971). *A wakeful hypometabolic physiological state*. American Journal of Physiology. 221 (3) 795-799.

Wallace R.K and Benson H. (1972). *The physiological effects of Transcendental Meditation.* Scientific American 226:84-90.

Wasling P. (2021). *Knaster: medvetandets gåta och illusionen av ett jag.* Förlag: Volante.

Wechler D. (1939). *The measurement of adult intelligence.* Förlag: Balyimore Williams & Wilkins.

Wigan A. (2007). *A New View of Insanity.* Förlag: Kessinger publishing co

Wittgenstein L. (2015) *Traktatus Logico-Philosophicus.* Förlag:Routledge

Wittgenstein L. (2021). *Filosofiska undersökingar.* Bokförlaget Thales.